PRACTICAL APPLICATIONS OF PHYSICAL CHEMISTRY IN FOOD SCIENCE AND TECHNOLOGY

Innovations in Physical Chemistry: Monograph Series

PRACTICAL APPLICATIONS OF PHYSICAL CHEMISTRY IN FOOD SCIENCE AND TECHNOLOGY

Edited by
Cristóbal Noé Aguilar, PhD
José Sandoval-Cortés, PhD
Juan Alberto Ascacio Valdes, PhD
A. K. Haghi, PhD

AAP | APPLE ACADEMIC PRESS

First edition published [2021]

Apple Academic Press Inc.
1265 Goldenrod Circle, NE,
Palm Bay, FL 32905 USA

4164 Lakeshore Road, Burlington,
ON, L7L 1A4 Canada

CRC Press
6000 Broken Sound Parkway NW,
Suite 300, Boca Raton, FL 33487-2742 USA

2 Park Square, Milton Park,
Abingdon, Oxon, OX14 4RN UK

Apple Academic Press exclusively co-publishes with CRC Press, an imprint of Taylor & Francis Group, LLC

Library and Archives Canada Cataloguing in Publication

Title: Practical applications of physical chemistry in food science and technology / edited by Cristóbal Noé Aguilar, PhD, José Sandoval-Cortés, PhD, Juan Alberto Ascacio Valdes, PhD, A.K. Haghi, PhD.

Names: Aguilar, Cristóbal Noé, editor. | Cortés, José Sandoval, editor. | Ascacio-Valdes, Juan Alberto, editor. | Haghi, A. K., editor.

Series: Innovations in physical chemistry.

Description: Series statement: Innovations in physical chemistry: monographic series | Includes bibliographical references and index.

Identifiers: Canadiana (print) 20200293109 | Canadiana (ebook) 2020029332X | ISBN 9781771888943 (hardcover) | ISBN 9781003020004 (eBook)

Subjects: LCSH: Food industry and trade. | LCSH: Food science. | LCSH: Chemistry, Physical and theoretical. | LCSH: Food—Analysis.

Classification: LCC TP372.5 .P73 2021 | DDC 664—dc23

Library of Congress Cataloging-in-Publication Data

CIP data on file with US Library of Congress

ISBN: 978-1-77188-894-3 (hbk)
ISBN: 978-1-00302-000-4 (ebk)

ABOUT THE EDITORS

Cristóbal Noé Aguilar, PhD

Cristóbal N. Aguilar, Chemist, PhD, is the Director of the Research and Postgraduate Programs at Autonomous University of Coahuila (2018–2021), Mexico, where from 2014 to 2018, he was the Dean of the School of Chemistry. He is a Level III member of S.N.I. (Mexican System of Researchers). He has received several prizes and awards, among which the most important are the National Prize of Research 2010 of the Mexican Academy of Sciences, the Prize "Carlos Casas Campillo 2008" of the Mexican Society of Biotechnology and Bioenegineering, the National Prize AgroBio–2005, the Mexican Prize in Food Science and Technology from CONACYT-Coca Cola México 2003, the 2018 Outstanding Researcher Award by the International Association of Bioprocessing, and the 2019 Coahuila State Innovation Science and Technology Award. He has developed more than 30 research projects, including seven international exchange projects (including ECOS–Nord, Alpha Network–EU Project, and Marie Curie EU Project). He has been an advisor of 20 PhD theses, 40 MSc theses, and 50 BSc theses. He is a member of the Mexican Academy of Science (since 2014).

José Sandoval-Cortés, PhD

José Sandoval Cortes, Chemist, PhD, is a full-time Professor at the School of Chemistry, Autonomous University of Coahuila, Mexico, for the last 12 years. He worked in a research stay in electrochemistry at the University of Paris 6 Pierre et Marie Curie, France; the National University of Cordoba, Argentina; and the National University of Rio Cuarto, Argentina. His research work has been published in six papers in indexed journals, three book chapters, and 20 contributions at scientific meetings. Since 2018, he has been part of the administrative staff of the Research and Postgraduate Studies Office at the Autonomous University of Coahuila. He received the Pharmaceutical Chemist Biologist degree in 2001, the MSc degree in chemistry in 2003, and the PhD degree in electrochemistry in 2006 from the University of Guanajuato, Mexico.

Juan Alberto Ascacio Valdes, PhD

Juan Alberto Asacio Valdes, Chemist, PhD, is a full-time Professor at the School of Chemistry, Autonomous University of Coahuila, Mexico, for the last four years. He worked in research stays in food research, bioprocesses, bioactive compounds, and nutraceuticals at CEBAS-CSIC, Murcia, Spain (2012) and at Jiangnan University, Wuxi, China (2017); and had a postdoctoral stay at Autonomous Agrarian University "Antonio Narroz", Mexico (2013–2015). His work has been published in 36 papers in indexed journals, 15 book chapters, and 30 contributions at scientific meetings. Since 2008, he has been a member of the Mexican Society of Biotechnology. He received the Pharmaceutical Chemist Biologist degree in 2006 from the Autonomous University of Coahuila, specializing in microbiology, and the MSc and PhD degrees in food science and technology from the same university in 2012.

A. K. Haghi, PhD

A. K. Haghi, PhD, is the author and editor of over 200 books, as well as over 1000 published papers in various journals and conference proceedings. Dr. Haghi has received several grants, consulted for a number of major corporations, and is a frequent speaker to national and international audiences. Since 1983, he served as professor at several universities. He is the former Editor-in-Chief of the *International Journal of Chemoinformatics and Chemical Engineering* and *Polymers Research Journal* and is on the editorial boards of many international journals. He is also a member of the Canadian Research and Development Center of Sciences and Cultures (CRDCSC), Montreal, Quebec, Canada.

INNOVATIONS IN PHYSICAL CHEMISTRY: MONOGRAPH SERIES

This book series offers a comprehensive collection of books on physical principles and mathematical techniques for majors, non-majors, and chemical engineers. Because there are many exciting new areas of research involving computational chemistry, nanomaterials, smart materials, high-performance materials, and applications of the recently discovered graphene, there can be no doubt that physical chemistry is a vitally important field. Physical chemistry is considered a daunting branch of chemistry—it is grounded in physics and mathematics and draws on quantum mechanics, thermodynamics, and statistical thermodynamics.

Editors-in-Chief

A. K. Haghi, PhD
Former Editor-in-Chief, *International Journal of Chemoinformatics* and *Chemical Engineering and Polymers Research Journal*; Member, Canadian Research and Development Center of Sciences and Cultures (CRDCSC), Montreal, Quebec, Canada
E-mail: AKHaghi@Yahoo.com

Lionello Pogliani, PhD
University of Valencia-Burjassot, Spain
E-mail: lionello.pogliani@uv.es

Ana Cristina Faria Ribeiro, PhD
Researcher, Department of Chemistry, University of Coimbra, Portugal
E-mail: anacfrib@ci.uc.pt

BOOKS IN THE SERIES

- **Applied Physical Chemistry with Multidisciplinary Approaches**
  Editors: A. K. Haghi, PhD, Devrim Balköse, PhD, and
  Sabu Thomas, PhD

- **Biochemistry, Biophysics, and Molecular Chemistry: Applied Research and Interactions**
  Editors: Francisco Torrens, PhD, Debarshi Kar Mahapatra, PhD, and A. K. Haghi, PhD
- **Chemistry and Industrial Techniques for Chemical Engineers**
  Editors: Lionello Pogliani, PhD, Suresh C. Ameta, PhD, and A. K. Haghi, PhD
- **Chemistry and Chemical Engineering for Sustainable Development: Best Practices and Research Directions**
  Editors: Miguel A. Esteso, PhD, Ana Cristina Faria Ribeiro, and A. K. Haghi, PhD
- **Chemical Technology and Informatics in Chemistry with Applications**
  Editors: Alexander V. Vakhrushev, DSc, Omari V. Mukbaniani, DSc, and Heru Susanto, PhD
- **Engineering Technologies for Renewable and Recyclable Materials: Physical–Chemical Properties and Functional Aspects**
  Editors: Jithin Joy, Maciej Jaroszewski, PhD, Praveen K. M., and Sabu Thomas, PhD, and Reza Haghi, PhD
- **Engineering Technology and Industrial Chemistry with Applications**
  Editors: Reza Haghi, PhD, and Francisco Torrens, PhD
- **High-Performance Materials and Engineered Chemistry**
  Editors: Francisco Torrens, PhD, Devrim Balköse, PhD, and Sabu Thomas, PhD
- **Methodologies and Applications for Analytical and Physical Chemistry**
  Editors: A. K. Haghi, PhD, Sabu Thomas, PhD, Sukanchan Palit, and Priyanka Main
- **Modern Green Chemistry and Heterocyclic Compounds: Molecular Design, Synthesis, and Biological Evaluation**
  Editors: Ravindra S. Shinde, and A. K. Haghi, PhD
- **Modern Physical Chemistry: Engineering Models, Materials, and Methods with Applications**
  Editors: Reza Haghi, PhD, Emili Besalú, PhD, Maciej Jaroszewski, PhD, Sabu Thomas, PhD, and Praveen K. M.

- **Molecular Chemistry and Biomolecular Engineering: Integrating Theory and Research with Practice**
  Editors: Lionello Pogliani, PhD, Francisco Torrens, PhD, and A. K. Haghi, PhD
- **Physical Chemistry for Chemists and Chemical Engineers: Multidisciplinary Research Perspectives**
  Editors: Alexander V. Vakhrushev, DSc, Reza Haghi, PhD, and J. V. de Julián-Ortiz, PhD
- **Physical Chemistry for Engineering and Applied Sciences: Theoretical and Methodological Implication**
  Editors: A. K. Haghi, PhD, Cristóbal Noé Aguilar, PhD, Sabu Thomas, PhD, and Praveen K. M.
- **Practical Applications of Physical Chemistry in Food Science and Technology**
  Editors: Cristóbal Noé Aguilar, PhD, José Sandoval-Cortés, PhD, Juan Alberto Ascacio Valdes, PhD, and A. K. Haghi, PhD
- **Research Methodologies and Practical Applications of Chemistry**
  Editors: Lionello Pogliani, PhD, A. K. Haghi, PhD, and Nazmul Islam, PhD
- **Theoretical Models and Experimental Approaches in Physical Chemistry: Research Methodology and Practical Methods**
  Editors: A. K. Haghi, PhD, Sabu Thomas, PhD, Praveen K. M., and Avinash R. Pai

CONTENTS

Contributors .... *xiii*

Abbreviations .... *xvii*

Preface .... *xix*

1. **Physical Chemistry on Food Science and Technology 1**
   Cristian Hernández-Hernández, F. Javier Alonso-Montemayor, J. A. Ascacio-Valdés, J. Sandoval-Cortés, C. N. Aguilar, and A. K. Haghi

2. **Physicochemical Properties and Extraction Methodologies of Agroindustrial Wastes to Produce Bioactive Compounds 39**
   Maricela Esmeralda-Guzmán, Catalina de Jesús Hernández-Torres, Desiree Dávila-Medina, Thelma Morales-Martínez, and Leonardo Sepúlveda

3. **Physicochemical Properties of Products Made from Mixtures of Corn and Legumes 61**
   Daniela Sánchez-Aldana Villarruel, Martha Yareli Leal Ramos, Tomás Galicia García, Ruben Marquez Meléndez, and Ricardo Talamás Abbud

4. **Analysis of Physicochemical and Nutritional Properties of Rambutan (*Nephelium lappaceum* L.) Fruit 95**
   José C. De León-Medina, Cristian Hernández-Hernández, Leonardo Sepúlveda, Adriana C. Flores-Gallegos, José Sandoval-Cortés, Jose J. Buenrostro-Figueroa, Cristóbal N. Aguilar, and Juan A. Ascacio-Valdés

5. **Development of Modern Electroanalytical Techniques Based on Electrochemical Sensors and Biosensors to Quantify Substances of Interest in Food Science and Technology 109**
   María A. Zon, Fernando J. Arévalo, Adrian M. Granero, Sebastián N. Robledo, Gastón D. Pierini, Walter I. Riberi, Jimena C. López, and Héctor Fernández

6. **Pomegranate Seeds as a Potential Source of Punicic Acid: Extraction and Nutraceutical Benefits 129**
   Juan M. Tirado-Gallegos, R. Baeza-Jiménez, Juan A. Ascacio-Valdés, Juan C. Bustillos-Rodríguez, and Juan Buenrostro-Figueroa

7. **Nutritional Evaluation of Waste in the Citrus Industry 155**
   Patricia M. Albarracín, María F. Lencina, and Norma G. Barnes

8. Microencapsulation as a Technological Alternative in the Food Industry for Conservation of Betalaines 161

Juan Antonio Ugalde-Medellín, Lluvia Itzel López-López, Juan Guzmán-Ceferino, Sonia Yesenia Silva-Belmares, Cristóbal Noé Aguilar, and Janeth Margarita Ventura-Sobrevilla

9. Nanoemulsions for Edible Coatings: Stabilizing and Bioactive Properties 183

Miguel De León-Zapata, Lorenzo Pastrana-Castro, Letricia Barbosa-Pereira, María L. Rua-Rodríguez, Janeth Ventura, Thalia Salinas, Raul Rodríguez, Juan A. Ascacio-Valdés, José Sandoval-Cortés, and Cristóbal N. Aguilar

10. Enhancing the Added Value of Sorghum by Biomolecule Content and Bioprocessing.. 199

Marisol Cruz-Requena, Leopoldo J. Ríos-González, José Antonio De La Garza-Rodríguez, Sócrates Palacios Ponce, and Miguel A. Medina-Morales

11. Efficiency of Fertilizer Application on Spring Wheat in the Conditions of the Ural Region 223

Ekaterina A. Semenova, Rafail A. Afanas'ev, and Michael Smirnov

Index.. 243

CONTRIBUTORS

Ricardo Talamás Abbud
Facultad de Universidad Autónoma de Chihuahua, Circuito Universitario S/N, Campus Uach II, Chihuahua 31125, Chihuahua, México

Rafail A. Afanas'ev
Pryanishnikov All-Russian Scientific Research Institute of Agrochemistry, d. 31A, Pryanishnikova Street, Moscow 127550, Russia

Cristóbal N. Aguilar
Bioprocesses & Bioproducts Group, Food Research Department, School of Chemistry, Autonomous University of Coahuila, Saltillo 25280, Coahuila, México

Patricia M. Albarracín
Process Engineering and Industrial Management Department, Faculty of Exact Sciences and Technology, National University of Tucumán, Argentina

F. Javier Alonso-Montemayor
Bioprocesses & Bioproducts Group, Food Research Department, School of Chemistry, Autonomous University of Coahuila, Saltillo 25280, Coahuila, México

Fernando J. Arévalo
Grupo de Electroanalítica (GEANA), Departamento de Química, Facultad de Ciencias Exactas, Físico-Químicas y Naturales, Universidad Nacional de Río Cuarto, Agencia Postal No. 3, 5800 Río Cuarto, Argentina

Juan A. Ascacio-Valdés
Bioprocesses & Bioproducts Group, Food Research Department, School of Chemistry, Autonomous University of Coahuila, Saltillo 25280, Coahuila, México

R. Baeza-Jiménez
Research Center in Food and Development, A.C. 33089 Cd. Delicias, Chihuahua, México

Letricia Barbosa-Pereira
Department of Agriculture, Forest and Food Science, School of Science, University of Turin, Turin 10095, Italy

Norma G. Barnes
Process Engineering and Industrial Management Department, Faculty of Exact Sciences and Technology, National University of Tucumán, Argentina

Jose J. Buenrostro-Figueroa
Research Center for Food and Development A.C., 33088 Cd. Delicias, Chihuahua, México

Juan C. Bustillos-Rodríguez
Research Center in Food and Development, A.C. 31570 Cd. Cuauhtémoc, Chihuahua, México

Marisol Cruz-Requena
Food Research Department, School of Chemistry, Autonomous University of Coahuila, Saltillo 25280, Coahuila, México

Desiree Dávila-Medina
Group of Bioprocess and Microbial Biochemistry, School of Chemistry, Autonomous University of Coahuila, Saltillo, Coahuila, México

Catalina de Jesús Hernández-Torres
Group of Bioprocess and Microbial Biochemistry, School of Chemistry, Autonomous University of Coahuila, Saltillo, Coahuila, México

José Antonio De La Garza-Rodríguez
Biotechnology Department, School of Chemistry, Autonomous University of Coahuila, Saltillo 25280, Coahuila, México

Maricela Esmeralda-Guzmán
Group of Bioprocess and Microbial Biochemistry, School of Chemistry, Autonomous University of Coahuila, Saltillo, Coahuila, México

Héctor Fernández
Grupo de Electroanalítica (GEANA), Departamento de Química, Facultad de Ciencias Exactas, Físico-Químicas y Naturales, Universidad Nacional de Río Cuarto, Agencia Postal No. 3, 5800 Río Cuarto, Argentina

Adriana C. Flores-Gallegos
Bioprocesses & Bioproducts Group, Food Research Department, School of Chemistry, Autonomous University of Coahuila, Saltillo 25280, Coahuila, México

Tomás Galicia García
Facultad de Universidad Autónoma de Chihuahua, Circuito Universitario S/N, Campus Uach II, Chihuahua 31125, Chih., México

Adrian M. Granero
Grupo de Electroanalítica (GEANA), Departamento de Química, Facultad de Ciencias Exactas, Físico-Químicas y Naturales, Universidad Nacional de Río Cuarto, Agencia Postal No. 3, 5800 Río Cuarto, Argentina

Juan Guzmán-Ceferino
Agricultural Sciences Academic Division, Juárez Autonomous University of Tabasco, Tabasco, Mexico

A. K. Haghi
Canadian Research and Development Center of Sciences and Cultures (CRDCSC), Montreal, Quebec, Canada

Cristian Hernández-Hernández
Bioprocesses & Bioproducts Group, Food Research Department, School of Chemistry, Autonomous University of Coahuila, Saltillo 25280, Coahuila, México

María F. Lencina
Process Engineering and Industrial Management Department, Faculty of Exact Sciences and Technology, National University of Tucumán, Argentina

José C. De León-Medina
Bioprocesses & Bioproducts Group, Food Research Department, School of Chemistry, Autonomous University of Coahuila, Saltillo 25280, Coahuila, México

Miguel De León-Zapata
Department of Research (DIA-UAdeC), School of Chemistry, University Autonomous of Coahuila. Saltillo, 25280 Coahuila, México

Jimena C. López
Grupo de Electroanalítica (GEANA), Departamento de Química, Facultad de Ciencias Exactas, Físico-Químicas y Naturales, Universidad Nacional de Río Cuarto, Agencia Postal No. 3, 5800 Río Cuarto, Argentina

Lluvia Itzel López-López
Institute of Research in Desert Areas, Autonomous University of San Luis Potosi, San Luis Potosi 78377, Mexico

Ruben Marquez Meléndez
Facultad de Universidad Autónoma de Chihuahua, Circuito Universitario S/N, Campus Uach II, Chihuahua 31125, Chih., México

Miguel A. Medina-Morales
Biotechnology Department, School of Chemistry, Autonomous University of Coahuila, Saltillo 25280, Coahuila, México

Thelma Morales-Martínez
Group of Bioprocess and Microbial Biochemistry, School of Chemistry, Autonomous University of Coahuila, Saltillo, Coahuila, México

Sócrates Palacios Ponce
Mechanical Engineering and Production Sciences School, ESPOL Polytechnic University (Litoral), Campus Gustavo Galindo, CP 09-01-5683 Guayaquil, Ecuador

Lorenzo Pastrana-Castro
Health, Food and Environment Department, International Iberian Nanotechnology Laboratory (INL), Braga 4715-330, Braga, Portugal

Gastón D. Pierini
Grupo de Electroanalítica (GEANA), Departamento de Química, Facultad de Ciencias Exactas, Físico-Químicas y Naturales, Universidad Nacional de Río Cuarto, Agencia Postal No. 3, 5800 Río Cuarto, Argentina

Martha Yareli Leal Ramos
Facultad de Universidad Autónoma de Chihuahua, Circuito Universitario S/N, Campus Uach II, Chihuahua 31125, Chih., México

Walter I. Riberi
Grupo de Electroanalítica (GEANA), Departamento de Química, Facultad de Ciencias Exactas, Físico-Químicas y Naturales, Universidad Nacional de Río Cuarto, Agencia Postal No. 3, 5800 Río Cuarto, Argentina

Leopoldo J. Ríos-González
Biotechnology Department, School of Chemistry, Autonomous University of Coahuila, Saltillo 25280, Coahuila, México

Sebastián N. Robledo
Departamento de Tecnología Química, Facultad de Ingeniería, Universidad Nacional de Río Cuarto, Agencia Postal No. 3, 5800 Río Cuarto, Argentina

Raul Rodríguez
Department of Research (DIA-UAdeC), School of Chemistry, University Autonomous of Coahuila. Saltillo, 25280 Coahuila, México

María L. Rua-Rodríguez
Laboratory of Analytical and Biochemistry Food, School of Science, University of Vigo, Ourense 32004, Galicia, España

Thalia Salinas
Department of Research (DIA-UAdeC), School of Chemistry, University Autonomous of Coahuila. Saltillo, 25280 Coahuila, México

José Sandoval-Cortés
Analytical Chemistry Department, School of Chemistry, Autonomous University of Coahuila, Saltillo 25280, Coahuila, México
Canadian Research and Development Center of Sciences and Cultures (CRDCSC), Montreal, Quebec, Canada

Ekaterina A. Semenova
Agrochemical Center of the Sverdlovsk, d. 109, Furmanova Street, Ekaterinburg 620144, Russia

Leonardo Sepúlveda
Group of Bioprocess and Microbial Biochemistry, School of Chemistry, Autonomous University of Coahuila, Saltillo, Coahuila, México
Bioprocesses & Bioproducts Group, Food Research Department, School of Chemistry, Autonomous University of Coahuila, Saltillo 25280, Coahuila, México

Sonia Yesenia Silva-Belmares
School of Chemistry, Autonomous University of Coahuila. Saltillo, Coahuila 25280, México

Michael Smirnov
Agrochemical Center of the Sverdlovsk, d. 109, Furmanova Street, Ekaterinburg 620144, Russia

Juan M. Tirado-Gallegos
School of Animal Sciences and Ecology, Autonomous University of Chihuahua, 31453 Chihuahua, Chihuahua, México

Juan Antonio Ugalde-Medellín
School of Chemistry, Autonomous University of Coahuila. Saltillo, Coahuila 25280, México

Juan Alberto Ascacio Valdes
School of Chemistry, Universidad Autónoma de Coahuila, México

Janeth Margarita Ventura-Sobrevilla
School of Nutrition, Autonomous University of Coahuila. Piedras Negras, Coahuila 26090, México

Janeth Ventura
Department of Research (DIA-UAdeC), School of Chemistry, University Autonomous of Coahuila. Saltillo, 25280 Coahuila, México

Daniela Sánchez-Aldana Villarruel
Facultad de Universidad Autónoma de Chihuahua, Circuito Universitario S/N, Campus Uach II, Chihuahua 31125, Chih., México

María A. Zon
Grupo de Electroanalítica (GEANA), Departamento de Química, Facultad de Ciencias Exactas, Físico-Químicas y Naturales, Universidad Nacional de Río Cuarto, Agencia Postal No. 3, 5800 Río Cuarto, Argentina

ABBREVIATIONS

BUT	butein
BHA	butylated hydroxyanisole
BHT	butylated hydroxytoluene
CMC	carboxymethyl cellulose
CIT	citrinin
CLnA	conjugated linolenic acid
CL	cooking loss
CDG	cyclo-dopa-5-*O*-glycoside
DON	deoxinivalenol
DM	dried matter
ERH	equilibrium relative humidity
EUG	eugenol
FIS	fisetin
FDA	Food and Drug Administration
GSS	global score
HRP	horseradish peroxidase
HI	hydrothermal index
LUT	luteolin
MON	moniliformin
MO	morin
OTA	ochratoxin A
OCT	optimum cooking time
PAT	patulin
PSO	pomegranate seed oil
PG	propyl gallate
PuA	punicic acid
RUT	rutin
SEE	snacks by extrusion
SFC	solid fat content
SPE	soybean peroxidase enzyme
SWV	square wave voltammetry
STEH	sterigmatocystin (STEH)
TBHQ	*tert*-butyl hydroxyquinone

TOM	total organic matter
t-RES	trans-resveratrol
UMEs	ultramicroelectrodes
UAE	ultrasound-assisted extraction
WA	water absorption
WAC	water-absorbing capacity
WF	wheat flour
ZEA	zearalenone

PREFACE

The book provides comprehensive information, original research contributions, and scientific advances in the field of practical applications of physical chemistry in food science and technology. The chapters cover broad research areas offering original and novel highlights in research studies of basic and applied physical chemistry in the field of foods and biological systems and enhance the exchange of scientific literature.

This book covers the basic requirements of students, teachers, and researchers interested in the physical chemistry of foods. The editors have shared efforts on the topic because they are members of the School of Chemistry at the Autonomous University of Coahuila (Mexico) and they have experience in teaching and research.

Having in mind that physical chemistry is considered a daunting branch of chemistry, the book offers a comprehensive collection of chapters on physical principles and modern techniques for majors, nonmajors, and chemical engineers.

The book explores the most important advances in the practical applications of physical chemistry in food science and technology, with special emphasis on the challenges faced by this sector in the era of sustainable development and by the food and agricultural industries in a frame of requirements for sustainable development goals. The volume provides a detailed and up-to-date revision of useful tools in teaching and research related to food science, technology, and engineering. This book demonstrates the potential and actual developments across the innovative advances in the design and development of physical chemistry strategies and tools forfood science and technology.

The book comprises 11 chapters, which include relevant information about concepts, properties and analyses, electrochemistry, energy, emerging technologies, valorization of food residues, bioactive compounds and bioactivities, separation extraction, microencapsulation, and nanoemulsions. Due to the book's qualified and innovative content, there is no doubt that it will be an important reference for food science and technology research, for undergraduate and graduate students, and for professionals in the food industry.

In general, physical chemistry is a multidisciplinary science that involves the knowledge of thermodynamics, quantum chemistry, transport phenomena, electrochemistry, among others areas. Physical chemistry is useful to understand the changes in foods and provide the basis for food processing optimization. In this context, it takes into account the thermodynamic models, and food scientists and engineers can develop new or better techniques for the conservation and exploitation of food. In the first chapter, the basic concepts of physical chemistry are briefly explained along with examples of applications in food science and technology.

Currently, the food industry generates a large amount of agroindustrial waste that seriously affects the environment. This problem is spreading slowly around the world due to the lack of regulations and the inadequate treatment of these agroindustrial wastes. The interest of modern biotechnology is to use agroindustrial waste to produce bioactive compounds that are of high added value and different uses in the industry. Bioactive compounds or secondary metabolites from plant origin, also called phytochemicals, are compounds with important biological properties and human health benefits, such as antimicrobial, antiviral, antioxidant, anticancer, among others. Chapter 2 explains topics related to the physicochemical properties of agroindustrial wastes that are used in the north of México to produce bioactive compounds. In addition, some characteristics of wastes of nutshell (*Juglans regia*), grape (*Vitis vinífera*), and nopal (*Opuntia ficus-indica*) are described. Finally, some bioactive compounds can be obtained by emerging technologies used in the food industry.

Physicochemical properties play an important role in maintaining the quality of the products made from mixtures of corn and legumes that are fundamental in the research and development of new products, the design of equipment, the improvement of processes, and the quality control of raw materials, intermediates, and finished products. Physicochemical properties are intimately related to the functional properties of the constituents of the food system, as well as the operational variables that are applied in the different stages of the process. Likewise, nixtamalization of maize is a thermoalkaline treatment that is responsible for important changes in the physicochemical, thermal, nutritional, and sensory characteristics in flours, doughs, and finished products. In Chapter 3, the physicochemical properties of different products made of mixtures of corn and legume products such as flours, doughs, bakery, tortillas, snacks, and breakfast cereals is reviewed.

Rambutan is an exotic fruit native to Southeast Asia. It is composed of peel, pulp, and seed; the pulp is the only edible part of the fruit. The cultivars of rambutan have been expanded in many parts of the world, and its consumption is principally fresh, but it is also processed in the industry for the production of juices, jams, or jellies, leaving the peel and seeds as byproducts with potential applications in the industry due to their physicochemical properties, especially the seeds. The peel has been reported as an interesting source of bioactive compounds and a source of lignocellulosic material that can be used for applications such as bionanocomposites. The rambutan seed is the part of the fruit where most physicochemical properties are reported, which are attributed to its high amount of fat and starch. The physicochemical properties of rambutan seed fat and starch can be compared with those of the other commercial raw materials used widely in industry, like cocoa butter or corn starch. So, rambutan seeds seem to be a potential new source for use as a raw material in the pharmaceutical, cosmetic, and food industries. So, Chapter 4 summarizes all of the physicochemical and nutritional properties that have been reported for the rambutan pulp, peel, and seed and the potential applications of these properties in the industry.

The importance of electroanalytical techniques in the determination of many substances in the science and technology of food has increased significantly in recent years. This has been possible thanks to the great advances made in the fields of electronics, biotechnology, the development of new nanomaterials, etc. They have allowed, in turn, notable advances in the development of modern electroanalytical techniques. Some of the advantages of these techniques compared to the conventional analytical techniques are higher speed, greater simplicity, lower cost, lower consumption of solvents, and the possibility of making determinations in real samples without any pretreatment. Chapter 5 discusses the development of electrochemical sensors and biosensors for the quantification of important components of foods (natural and from contamination), such as mycotoxins, synthetic and natural antioxidants, metals, etc., performed by the members of the GEANA** group over the last 20 years. This chapter highlights the importance of combining the results of basic research, from the point of view of chemical physics, with the advantageous application of such results to the development of electroanalytical methodologies in quality control systems for food and other important areas.

Pomegranate (*Punica granatum* L.) is a fruit cultivated under diverse climatic conditions, being an edible fruit with great adaptability and flexibility around the world. Its consumption has been related to health benefits. Pomegranate seeds are an interesting component of this fruit due to their oil content (up to 50 wt.%), which has been distinguished for its pharmaceutical applications and nutraceutical properties. The main fatty acid residue identified in the oil is punicic acid (PuA) (≈80%, with respect to the total fatty acid content). PuA can exert important bioactivities such as anticancer, hypolipidemic, antidiabetes, antiobesity, antioxidant, anti-inflammatory, among others. One of the most potent sources of PuA is pomegranate seed, a byproduct obtained during the industrialization of this fruit. In this sense, one of the strengths in PuA research is the revalorization of this agroindustrial residue for the further recovery of this bioactive fatty acid. On the other hand, new and novel applications for PuA can be developed. Therefore, in Chapter 6, it will be fully detailed the research fields on PuA, its application in pharmaceutical, cosmetic, and food industries, the different existing sources, the extraction methods, the analytical techniques for its identification, and its nutraceutical benefits in food and human health.

In the citrus industry, as byproducts of the process to obtain citrus juices, there remain the crusts, membranes, part of the pulp, and seeds. Chapter 7 provides the nutritional study of wet husk, husk dust, residue from the crushing of the dry skin, and wet residue from the pulp and solids filtered from the liquid effluents of the plant. This study was carried out with the aim to analyze the use of these wastes in the elaboration of balanced feed for animals. Samples were taken from a local citrus industry. The contents of dry matter, lipids, fibers, proteins, carbohydrates, and flavonoids were determined in triplicate. The obtained values were analyzed statistically in Excel 2010 and showed a standard deviation on the order of 5%. The powder of lemon peel showed the highest content of dry matter (99.7%), followed by that of lipids, proteins, fibers, and carbohydrates. For the wet shell, the content was 37.7%, and for the filtered wet residues, the content was 23.5%. The obtained results allow affirming that the industrial waste analyzed can be used in animal balanced feed, making supplementation of the deficient nutrients.

Nowadays, it is widely recognized that consuming fruits and vegetables brings many benefits to human health due to their content of different biomolecules (phytochemicals). Betalains are water-soluble biomolecules derived from the condensation of betalamic acid with a primary or secondary amine. Due to differences in their structural configuration, betalains are divided into two groups: betacyanins and betaxanthins. Both groups show optical properties because they possess two chiral centers at C-2 and C-15 with conjugated double bonds and present the maximum light absorption at 480 nm for betaxanthins and at 536 nm for betacyanins. These compounds also present colors; this feature has led to their wide use in the food area as natural dyes, and even in the cosmetic industry. The synthetic dyes have been associated with a number of diseases and disorders; thus, the need for natural products to avoid adverse effects has raised. Betalains belong to a group of five natural coloring additives commonly used in the food industry and are used especially on meats, dairy products, dehydrated drinks, cold drinks, and jellies. The potential of betalains as food additives is potentialized when they are microencapsulated because this can prevent their degradation and maximize their store life. Therefore, in Chapter 8, it is highlighted that betalains are versatile molecules with applications in different areas from food to pharmacological industry, but the main focus is devoted to betalains as food additives for their staining capabilities, as well as antioxidant activity. However, their lability has been a problem that has diminished their implementation in recent years. For that reason, the implementation of technologies that prevent their degradation, like the microencapsulation process, is one of the most promising methods for the food industry because the color and the bioactivity are preserved, and more importantly, this is a cheap and efficient technique. In conclusion, the contributions to the mankind by the plants and their biomolecules are enormous and undeniable; however, it is necessary to study and find new compounds capable of overcoming the current problems as other molecules did in the past.

The aim of Chapter 9 was to evaluate the effect of extract of tarbush *Flourensia cernua* as a stabilizer, an antioxidant, and a fungistatic in a candelilla wax-based emulsion for edible coatings in refrigeration conditions for 7 weeks. The extract of tarbush was used as an active component of the emulsion. The extract of tarbush presented good antioxidant activity in oil-in-water emulsions, as measured by the inhibition of the hydroperoxide assay by conjugated dienes, ABTS, and DPPH. The results of the microbiological analysis demonstrated that the extract presented a higher fungistatic effect on yeast growth in the emulsion. The emulsion with the extract showed higher stability and antioxidant and fungistatic activity relative to those of the control, without any significant differences in the storage for 4 weeks in refrigeration conditions.

Sorghum is a relatively undervalued crop compared to corn and wheat. As it is a plant that thrives better in dry environments and has a versatile carbon fixation metabolism, it may represent an important underused source of high added-value chemicals and foods. By biotechnological processing, several outcomes may prove the said importance of sorghum. Most of the plant is useful for processing by biotechnological or food bioscience ways. For these purposes, the content of bioactive molecules, such as polyphenols as relative majority and other compounds such as stilbenoids. It is important to add that, for most of the native bioactive content of sorghum, physical processing may facilitate the assimilation of these molecules in the human diet with many beneficial effects. Also, the polysaccharide content can be processed for bioactive or high added-value molecule production by a biotechnological way. By enzymatic processing, the degree of polymerization of starch can be disrupted by liquefaction, and by microbial metabolism, several products may be produced. It is worth noting that the compounds mentioned in Chapter 10 have applications in many industrial areas.

The efficiency of fertilizer application on spring wheat in the conditions of the URAL region is presented in Chapter 11.

We are sure that the book will be of great interest and support for readers because it particularly focuses to meet the demands and needs of students, teachers, and researchers for the practical applications of food science and technology.

We deeply thank all the contributors who responded enthusiastically to the call issued by contributing original and novel documents. All the editors thank the contributing authors for their time and for sharing their knowledge for the benefits of students and professionals.

CHAPTER 1

PHYSICAL CHEMISTRY ON FOOD SCIENCE AND TECHNOLOGY

CRISTIAN HERNÁNDEZ-HERNÁNDEZ[1],
F. JAVIER ALONSO-MONTEMAYOR[1], JUAN A. ASCACIO-VALDÉS[1],
JOSÉ SANDOVAL-CORTÉS[2*], CRISTÓBAL NOÉ AGUILAR[1*], and
A.K. HAGHI[3]

[1]*Bioprocesses & Bioproducts Group, Food Research Department, School of Chemistry, Autonomus University of Coahuila, Saltillo 25280, Coahuila, México*

[2]*Canadian Research and Development Center of Sciences and Cultures (CRDCSC), Montreal, Quebec, Canada*

[3]*Canadian Research and Development Center of Sciences and Cultures (CRDCSC), Montreal, Quebec, Canada*

**Corresponding author. E-mail: josesandoval@uadec.edu.mx*

ABSTRACT

In general, physical chemistry is a multidisciplinary science that involves the knowledge of thermodynamics, quantum chemistry, transport phenomena, electrochemistry, among others. In foods, physical chemistry is useful to understand the changes in foods and provide the basis for food processing optimization. In this context, take in count thermodynamic models, food scientists and engineers can develop new or better techniques for the conservation and exploitation of food. In this chapter, the basic concepts of physical chemistry are briefly exposed with examples of their application in food science and technology.

1.1 INTRODUCTION

In general, science and technology are intimately related because science allows the development of new or enhanced technologies and methodologies. This also applies to food technology, which is a consequence of the scientific understanding of food as a phenomenon. In this context, food scientists want to describe and predict changes in food quality during their processing, storage, and handling. On the other hand, food technologists design and improve processes to provide specific properties to food (Walstra, 2003).

Some technological problems in foods demonstrate that the study of food as a phenomenon and the development of food technologies implies the knowledge of the physical fundamentals of the chemical systems in food. This is the definition of *physical chemistry*, which explores the chemical systems from the microscopic (molecules interactions) and macroscopic (large-scale properties) points of view (Levine, 2009).

Physical chemistry is divided into four areas: thermodynamics, quantum chemistry, statistical mechanics, and kinetics. *Thermodynamics* is the study of the interrelations of microscopical properties of a system in equilibrium. On the other hand, *quantum chemistry* is the result of applying the quantum physics of the atomic structure and molecular bonds. *Statistical mechanic* allows to calculate the macroscopical properties (thermodynamic) from molecular properties (quantum chemistry). Finally, *kinetics* describe and predict the chemical reaction rate based on thermodynamic, quantum chemistry, and statistical mechanic theories (Levine, 2009).

The physical chemistry aspects generally are important in food processing; however, the problems involved in food science and technology are more complex. In the first place, because foods have a wide composition, they are not in thermodynamic equilibrium. Also, several simultaneous chemical changes may occur. All of these issues make the application of physical chemistry theories difficult because food systems do not comply with the theory of basic assumptions. In the second place, most foods have two or more phases (are inhomogeneous systems) and other phenomena emerge such as *colloidal interactions* and *surface forces*. These phenomena occur on a larger scale than molecular but are not macroscopic properties; in other words, they occur on a mesoscopic scale (Walstra, 2003).

All these might lead to the opinion that applying physical chemistry to solve the problems of food science and technology is unfruitful. However, taking into account the basic scientific principles involved can be helpful indeed. In general, fundamental knowledge can be used to identify and explain the mechanism of the process and to found semi-quantitative relations; if theorizing is not based on scientific principles, it will often lead to wrong conclusions; and because foods are complex systems, they require widely different specifications, whose basic understanding and semi-quantitative relations should reduce greatly the number of trials (Walstra, 2003).

So, as an opinion, physical chemistry and the study of mesoscopic properties might be useful, in a complementary way, providing scientific perspectives to describe and explain the changes and processes involved in food as phenomena and food manufacture (Walstra, 2003).

1.1.1 ASPECTS OF THERMODYNAMICS

Thermodynamics arises from the search for solutions in the problems in the design of thermal machines. It is related to the exchange of energy between the components of a system and its environment. It studies natural phenomena and any system or device used by man. In addition, it is related to the field of transport phenomena, such as heat and material transfer. It also plays a very important role in food engineering, for example, in the control of air and moisture content in a final product. Several phase changes that occur in the storage or even processing of food are described through the principles of thermodynamics. The purpose of this chapter is to briefly introduce thermodynamic concepts and functions in food applications and to serve as a basis for the study and understanding of food engineering processes (Welti-Chanes et al., 2009).

What is thermodynamics? It is a set or part of the system that is conformed in the physical universe, submitted to diverse processes that imply exchanges of materials or energy in its environment. Also, it can be considered a set of useful mathematical relationships between quantities, in such a way that it can be used to quantify several quantities in an easier way, which allows us to measure many variables and unknowns (Jones, 1997).

The thermodynamic system is separated from the rest of the universe by limits or boundaries, subject to an open or closed system. An open system can exchange mass and energy with its environment; for example, you have food that is dehydrated in a hot air dryer. In this process, there is a mass flow and heat that passes through the food. On the other hand, a closed system cannot exchange mass. However, for energy as an example of this, we have hermetically packaged food that undergoes a process of sterilization or cooling. Finally, an isolated system that cannot exchange mass or energy constitutes an isolated system, so it is not common and is rarely useful (Jones, 1997).

Thermodynamics possesses a thermal property or variable of a system that mainly occupies enthalpy, entropy, and Gibbs free energy, as thermodynamic functions that can be expressed in terms of other variables, using mathematical transformations to calculate state function which is now possible using using enthalpy (*H*), entropy (*S*), and Gibbs free energy (*G*).

(Jones, 1997).

$$H = H(T, P) \tag{1.1}$$

$$S = S(T, P) \tag{1.2}$$

$$G = G(T, P) \tag{1.3}$$

1.1.2 ENTROPY

Entropy is also known as the nucleus of the second thermodynamic law because it is a measure of the amount of molecular disorder within a system and can increase or decrease the transport of energy across the system frontier (Dincer and Cengel, 2001).

The change in general entropy is positive. In the process of increasing or decreasing entropy from a warm body to a cold body, highly ordered structures are constructed from much simpler structures. Figure 1.1 shows the heat transfer process from an entropy point of view (Dincer and Cengel, 2001).

During the entropy process, a heat transfer process occurs and the net entropy increases. The increase in the entropy of a cold body compensates for the decrease in the entropy of a warm body. The processes usually only cause an entropy or molecular disorder (Dincer and Cengel, 2001).

Precisely, that is why the processes can only occur in the direction of a greater general entropy or molecular disorder, for this entropy is a measure of the disorder.

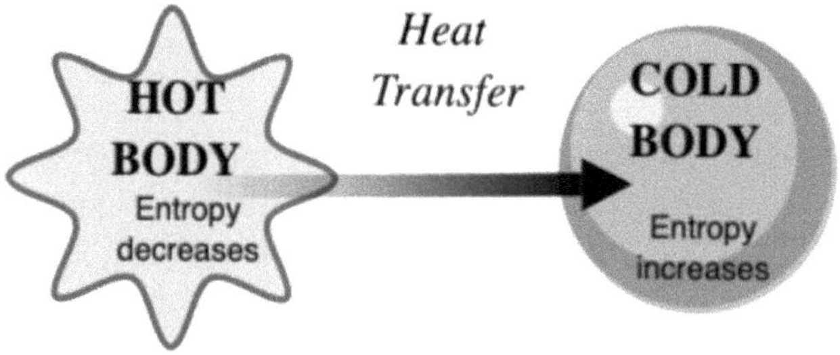

FIGURE 1.1 Increase and decrease of entropy in cold and hot bodies.

Entropy is given by

$$S = k_B \ln W \tag{1.4}$$

where k_B represents the Boltzmann constant (1.38×10^{-23}J/K) and W is the number of times the system can be organized (number of degrees of freedom).

1.1.3 ENTHALPY

Enthalpy is expressed as the sum of internal energy and extensive property and the product of the volume and pressure. Enthalpy depends on the temperature and sum of all chemical species of the flow and is considered as the total enthalpy of a flow (Singh and Heldman, 2009).

On the other hand, to estimate the energy of food through enthalpy, energy can be added, removed, or a change in temperature can be made. When changing the enthalpy temperature above the freezing point, the energy is sensitive, and below the freezing point, it not only consists sensitive energy but also latent energy. In general, the effect of the composition of the individual components of the enthalpies is small in most cases, and therefore, the partial molar enthalpy of any chemical species is based on the function of composition, pressure, and temperature (Singh and Heldman, 2009).

Therefore, enthalpy can be described using the following formula:

$$H = E_i + PV \tag{1.5}$$

The enthalpy H is represented in (kJ), P is the pressure (kPa), and V is the volume (m^3).

Otherwise, enthalpy can also be expressed in units of mass as follows:

$$H = E_i' + PV' \tag{1.6}$$

The enthalpy H is represented in the unit of mass (kJ/kg), E_i' is the internal energy per unit of mass (kJ/kg), and V' is the specific volume (m^3/kg).

Generally, the value of enthalpy is given by the relationship between the reference states where it is arbitrarily selected, that is, it is zero for translation (Singh and Heldman, 2009).

1.1.4 GIBBS ENERGY

Gibbs energy is a concept of free energy that was introduced by Willard Gibbs to give a criterion in the unidirectionality of spontaneous change. Now, when we talk about free energy, we will refer to Gibbs energy. Therefore, free energy is a property that can determine what will happen. For example, when we add sugar to water, it will dissolve and the sugar molecules will be distributed throughout the liquid (water), resulting in less free energy. In the example above, the increase in entropy has a greater effect than the increase in enthalpy (Jones, 1997). Because of this, each system will change to constant pressure until it has obtained the least energy from Gibbs, which is represented as follows:

$$G = H - TS \tag{1.7}$$

where G represents the Gibbs free energy (J/mol), H is the enthalpy (J/mol), T is the temperature (K), and S is the entropy (J/mol/K).

1.1.5 EQUATION OF STATE OF IDEAL GASES

The equation of state is the functional relationship between the properties of a system. In addition, thermodynamic properties are established in a simple system by any two independent properties (Singh and Heldman, 2009; Welti-Chanes et al., 2009).

Also, for a perfect equation of state in gaseous systems, the volume changes according to temperature, volume, and pressure (Welti-Chanes et al., 2009). This variation is almost independent of gas, so an equation of state has been proposed that is known as the ideal or perfect gas equation, and is expressed as follows:

$$PV' = RT_A \tag{1.8}$$

or

$$P = rRT_A \tag{1.9}$$

where P represents the absolute pressure (Pa), V' is the specific volume (m^3/kg), R is the gas constant (m^3 Pa/[kg K]), T_A is the absolute temperature (K), and r is the density (kg/m^3).

1.1.6 TRANSPORT PHENOMENA

All transport phenomena such as heat, mass, and fluid transfer (convection, conduction, and diffusion) are the result of an imbalance between the parts of the system. The principle of all of these are rigid by a universal law, and some more in-depth concepts of transport phenomena are needed to address the subject; however, they are only described in a general way in this chapter. It is also important to know that transport phenomena are often very much applied in the food industry, as they have an important link between material processing, quality, and product safety (Welti-Chanes et al., 2003).

1.1.7 RHEOLOGY REGIMES (FLOW AND VISCOSITY)

Rheology is the science that studies the relations between the force that acts on a material, its deformation, and its flow. Rheology can be used in solids or liquids as well; when a material is subjected to stress, it is deformed, and the nature of the deformation that occurs is a characteristic property of rheology (Walstra, 2003).

Usually, when there is a velocity gradient within the material, this indicates that it is a flow and is exhibited by the fluids. The importance of studying food rheology is to define a set of parameters that can be used to correlate with a quality attribute. This section deals with rheology in a general way, only dealing with the flow properties of fluids (Walstra, 2003).

1.1.8 FLOW

Generally, in a tension, fluid is applied and it will flow no matter how slight it may be. Therefore, a fluid can not only be a liquid but also gaseous;

however, it is mainly considered in liquids. Fluids can be divided into two types of flow regimes: laminar flow and turbulent flow (Berk, 2009). The variables of either of these two types, mass flow rate, fluid viscosity, density, and flow channel geometry, are combined in a dimensionless group also known as the Reynolds (Re) number as presented below:

$$\mathrm{Re} = \frac{Dv\rho}{\mu} \tag{1.10}$$

where D represents a linear dimension for the case of flow in a full diameter pipe (m), v is the mean linear velocity of the fluid (m/s), ρ is the density of the fluid (kg/m^3), and μ is the viscosity of the fluid (Pa s).

Finally, this combination of variables is represented by the Reynolds number, where the equilibrium between the inertia forces is represented in the equation in the nominator and the viscous restrictions are in the denominator (Berk, 2009).

1.1.9 LAMINAR FLOW

The laminar flow, also known as aerodynamic flow, is characterized by having movement in a single direction at all points of the fluid, that is, the trajectories of the elements in small volumes show a smooth and regular pattern. In the case of food processing, a laminar regime is common wherein speeds are relatively low and viscosities are relatively high (Berk, 2009).

1.1.9.1 LAMINAR FLOW IN A CYLINDER OR A TUBE

The fluid can be considered as the flow that flows at a certain radius, and certain parameters are taken, such as the distance from the central axis of the pipe, the constant linear velocity, the length of the layer, and the thickness. The expression of the shear rate as $-\mathrm{d}V/\mathrm{d}r$ satisfies the equality in the equation of the shear stress as a function of the shear rate because the shear stress is always positive. Therefore, fluids showing a linear increase in shear stress with a shear velocity are called Newtonian fluids, and the equation is represented as follows (Berk, 2009):

$$\tau = -\mu \frac{\mathrm{d}v}{\mathrm{d}r} \tag{1.11}$$

where τ is the shear stress, a term given to induced stress when molecules slide on top of each other along a defined plane, μ is the constant of proportionality or viscosity, and the velocity gradient dv/dr o γ is the shear rate at which one molecule slides over another.

1.1.10 TURBULENT FLOW

The turbulent flow is a type of flow where there is an implication in the element of volume so that it can move at any time in any direction and therefore is chaotic, even if the average flow is in one direction. In the turbulent flow, the speed increases and the flow lines become wavy forming eddies. According to the law of energy conservation, the sum of the kinetic energy of the flowing liquid and the potential energy must remain the same (Walstra, 2003).

For a flow to become turbulent, this must depend on the preponderance of the inertia stresses, that is, those proportional to the friction or viscosity stresses. Therefore, it is possible to have a proportional relation according to the number of Re without dimensions, as follows (Walstra, 2003):

$$\mathrm{Re} = \frac{Lv\rho}{\eta} \tag{1.12}$$

where L is the length, a characteristic perpendicular to the direction of flow, v is the average flow velocity, that is, the volumetric flow divided by the cross-sectional area of the flow channel, ρ is the mass density, and η is the dynamic shear viscosity.

1.1.10.1 TURBULENT FLOW IN A CYLINDER OR A TUBE

The turbulent flow has been found where turbulence occurs in a Reynolds number greater than 2000 and develops above 4000. Then, between a range of 2000 and 4000 is the transition where the flow can be laminar or turbulent (Welti-Chanes et al., 2005).

The most common calculation in a process concerning turbulent flow is the estimation of the pressure drop. This leads to an analysis of the definition of the dimensionless friction factor, as shown below (Welti-Chanes et al., 2005):

$$f = \frac{2\Delta PD}{Lv\rho} \tag{1.13}$$

where ΔP represents the pressure drop (Pa), D is the diameter of the channel (m), L is the longitude of the channel (m), v is the velocity of the medium (m/s), and r is the density of the fluid (kg/m^3).

1.1.11 VISCOSITY

Viscosity is defined as the measure to resist the flow of a fluid. Although molecules are known to be in constant random motion in a fluid, the net velocity in a particular direction is zero. To induce flow at a given velocity, the magnitude of the force is necessary since velocity is related to the viscosity of a fluid. When the molecules of a fluid slide on top of each other, there is a flow where the particular direction of these molecules is given in a given plane. Therefore, there must be a velocity gradient and a velocity difference between adjacent molecules (Toledo, 2007).

Figure 1.2 illustrates how viscosity works; a mass of fluid confined between two flat plates has been taken into account, where the bottom plate is held in place and the top plate moves at a constant speed in the x direction (Toledo, 2007).

Assuming that a liquid layer comes into contact with each plate and moves at the speed of that plate, it exerts an action called shear (Berk, 2009).

The shear force F_x is established by Newton's law and is necessary to keep the upper plate in motion as it is proportional to the area of the plate A and the velocity gradient dv_x/dz.

$$F = -\mu A \frac{dv}{dz} \tag{1.14}$$

In the viscosity, the shear force given by the shear force per unit area is shown, represented by the symbol τ. The cutting speed is represented by the speed gradient dv_x/dz, represented by the symbol γ, and the viscosity is represented by the symbol μ. So, the shear force can be represented as follows:

$$\tau = \mu\gamma$$

The viscosity is independent of the shear rate for many fluids. These fluids are referred to as Newtonian fluids. Water, milk, gases, and diluted solutions of solutes that have low molecular weights are Newtonian fluids. By contrast, other fluids such as concentrated suspensions and polymer solutions are not Newtonian fluids, as their visibility depends on the cutting speed (Berk, 2009; Toledo, 2007).

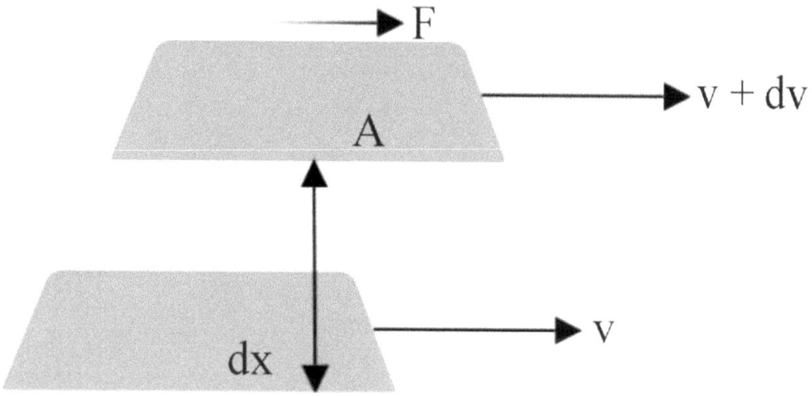

FIGURE 1.2 Function of viscosity.

1.1.12 WATER RELATIONS

Water is the most important element in food, as it has chemical and physical properties that make it useful for cells and organisms and is where most biochemical reactions take place. The water content varies widely in foods, so several of the properties of foods depend mostly on the water content. This water content affects the stability, safety, quality, and physical properties of foods (Holcroft, 2015).

Remember that water is a nutrient in food groups such as meats, grains, dairy products, vegetables, and fruits. Carbohydrates, minerals, proteins, and water-soluble vitamins are hydrophilic, while lipidic components are hydrophilic. Water concentrations in foods are very broad and range from 1% to 98%. In fresh and liquid products, we can find that they contain large amounts of water, as shown in Table 1.1, while dry and baked products contain little water. In most products, water is constitutive, but some food products may have water added during the process (Holcroft, 2015).

This is why the industry has long known how important it is to control free water in food. This is depicted by the way of expressing and providing important information about the quality of a product through the activity of water (a_w) because it also provides information about the possibility of growth of microbes on the surface. With this, we can have an indicator on the durability and stability of a sample (Holcroft, 2015).

TABLE 1.1 Water Content (%) of Some Vegetables and Fruits

Fruit	Water Content (%)	Vegetables	Water Content (%)
Apple	84	Asparagus	93
Avocado	76	Beans, green	89
Banana, green	76	Broccoli	90
Blueberry	83	Brussels sprouts	85
Cantaloupe	93	Cabbage	92
Cherry	80	Carrot	88
Citrus	89	Cauliflower	92
Grape	82	Lettuce	95
Grapefruit	89	Mushroom	91
Honeydew melon	93	Onion, dry	88
Kiwifruit	82	Pepper, sweet	92
Mango	82	Potato	78
Orange	86	Pumpkin	91
Peach	89	Spinach	93
Pear	83	Squash, summer	94
Plum	87	Squash, winter	85
Watermelon	93	Tomato, firm ripe	94

1.1.13 WATER ACTIVITY

Water activity (a_w) can be described as the relationship between the pure water vapor pressure in a solution and the water vapor pressure. Water activity can be measured by determining the equilibrium relative humidity (ERH) at any temperature and water content. Also, it is important to know that water activity is a measure of the moisture it finds free water available in a food product. Water activity is also used to estimate whether a food loses or absorbs water in a given environment. Finally, it could be said to be a parameter for assessing food quality (Sandulachi, 2013).

The state of water in a solid or solution is expressed by an activity coefficient, which is a thermodynamic measure of the chemical potential of water in the system (Sandulachi, 2013).

So, in the definition, it is the relationship between the vapor pressure of water in food (*P*) and the vapor pressure of pure water at the same temperature and total pressure (*Po*), as shown below:

$$a_w = \frac{P}{Po} \tag{1.15}$$

The determination of the ERH of a food product is defined as the relative humidity of the air surrounding the food in which the product is in equilibrium with the environment, that is, it does not lose its natural moisture. In this humid condition, no pathogenic or deteriorating microorganisms can develop, making it extremely important for food preservation (Kasaai, 2014; Sandulachi, 2013).

Therefore, there is a relationship between a_w and ERH, as both are related to vapor pressure.

$$a_w = \frac{\text{ERH}}{100} \tag{1.16}$$

1.1.14 SORPTION ISOTHERMS

Adsorption is a superficial phenomenon. Assuming that a liquid or a solid phase (the adsorbent) is in contact with another liquid phase. So now, molecules can be present in the fluid phase and can be adsorbed at the interface between the phases to form a generally monomolecular layer of the adsorbate. It should be remembered that the amount adsorbed is governed by the activity of the adsorbate and that for any combination of adsorbent, adsorbate, and temperature. It can be determined by an adsorption isotherm, that is, it can be visualized by means of a curve that is the result of the equilibrium ratio between the amount adsorbed per unit area and the activity of the adsorbate (Roos, 2003).

The absorption isotherm is the water content as a function of the equilibrium vapor pressure (P/P_o) at a constant temperature. The water activity and water content can be related by means of a graph, as illustrated in Figure 1.3, so that both parameters can be measured and the relationship of each parameter derived from the other can be defined (interpolation).

The use of the absorption isotherm may be a little practical, as not only does the relationship between humidity and a_w content changes with the measurement temperature but any variation in the composition of the material may also occur with a modifying effect (Roos, 2003).

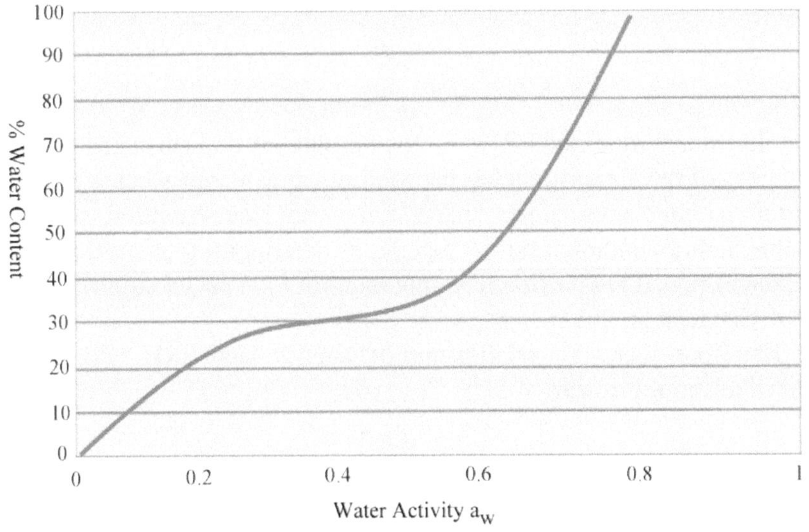

FIGURE 1.3 Sorption isotherm (illustration).

1.1.15 FORMATION OF EMULSIONS AND FOAMS

The emulsions and foams are found in systems that are often dispersed in food. To control the behavior of systems, it is essential to know the structures, the interfacial films, and the mechanical properties of the internal and external phases. In these systems, food macromolecules are widely used as functional ingredients in the stabilization and formation of these systems, such as polysaccharides and proteins. These molecules contain nonpolar and polar regions, which gives them surface-active properties. Also, oil droplets or gas bubbles are rapidly adsorbed and a film is formed on the surface during emulsification or foaming processes. Examples of food emulsions are meat products, margarine, frozen desserts, butter, and milk. These products contain water dispersed in oil or oil dispersed in water (Chiralt, 2009).

It is important to understand the complexity of food emulsions and foams to have better stability and maintenance in the quality of these types of food. In this chapter, the formation and stability of emulsions and foams are summarized in a general way (Chiralt, 2009).

1.1.15.1 EMULSIONS

An emulsion is a colloidal system that contains droplets of liquid dispersed in others, the two liquids being immiscible. Therefore, the droplets are in the dispersion phase and the liquid that contains them is called the continuous phase. Emulsions in food are present in two liquids, water and oil. Then, water is the continuous phase, where the emulsion is an oil-in-water (O/W), but if oil is the continuous phase, the emulsion is called water-in-oil (W/O), as illustrated in Figure 1.4. Oil and water emulsions are usually the most common, and frozen desserts, dressings, cake mix, mayonnaise, margarine, and butter are the examples of water-in-oil emulsions in food (Vaclavik and Christian, 2008).

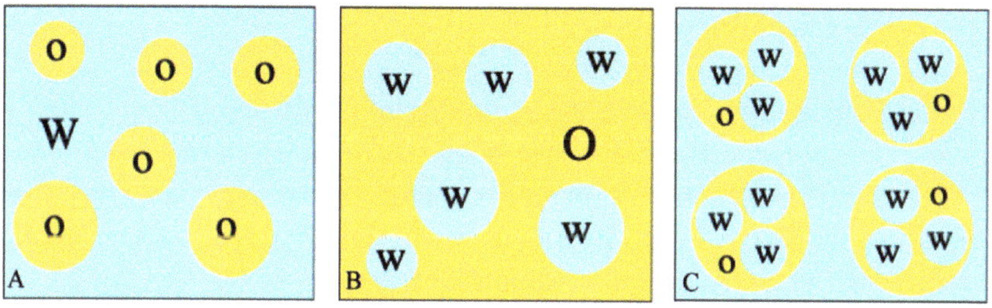

FIGURE 1.4 Emulsion of oil-in-water O/W (A), water-in-oil W/O (B), and multiple emulsions W/O/O (C).

It should also be noted that colloidal and liquid systems should not be taken literally because many food lipids can be partially crystallized; this will depend on the temperature of handling and consumption, for example, butter, whipped cream, and ice cream. In this case, the particles are dispersed in the food emulsions and can have different sizes and shapes, depending on the process conditions. Due to the size and surface of the droplets, emulsions are considered systems. Emulsifiers may become similar to colloidal dispersions or suns but with the exception that the dispersed phase is liquid and not solid in emulsions. An emulsion may also contain an emulsifier that covers the emulsion droplets and prevents them from reconverting or fusing (Chiralt, 2009; Vaclavik and Christian, 2008).

1.1.15.2 EMULSION STABILITY AND DESTABILIZATION

When the liquid phases are not miscible, emulsions can become potentially unstable and can lead to rapid phase separation if no kinetic factors are preventing them between the emulsions. Several factors destabilize emulsions such as range forces that exert an action for destabilization, namely, repulsive forces, gravitational forces, molecular forces, and flow forces. The action of these forces affect in different degrees and is responsible for the destabilization mechanisms (Walstra, 2003).

Primary processes can affect stability in emulsions, such as flocculation, cream, and coalescence, but sometimes Ostwald maturation and emulsion phase reversal is also considered (Walstra, 2003).

On the other hand, there are also principles for obtaining a stable emulsion. Therefore, this can be done by the following process:

1. Reduce the interfacial tension of each liquid by adding oil using the emulsifier method.
2. The mixture is then homogenized to supply energy.
3. The oily phase can be divided into drops of water.
4. The newly created oil drops adsorb the emulsifier on the surfaces.
5. Small drops will form and covered by an interfacial layer of emulsifier.
6. The oil interfacial size becomes larger.
7. Each drop of the aqueous phase will be surrounded by each drop of oil.
8. A thick emulsion will be obtained due to the presence of small drops of oil together with a thin continuous phase.
9. The emulsion will be stable when the interfacial film is strong.

1.1.16 FOAMS

A foam in its continuous liquid phase contains dispersed gas bubbles. A simple dispersion is made up of a liquid phase such as egg white, a clear example of diluted protein dispersion, which can be complex, containing ice crystals, emulsified fat droplets, and solid matter. One of the most important characteristics of foams is that they contribute to the texture and volume of many food products. They add volume to products such as ice cream and whipped cream and light texture to baked goods. One of the characteristics of unstable or poorly formed foams is that they result in low-volume (dense) products (Green et al., 2013).

Some of the factors that affect the stability or help stabilize the foam are summarized below:

Factors that help stabilize foams have

1. a stable viscoelastic surface film,
2. a very viscous continuous phase, and
3. low vapor pressure of a liquid.

Factors affecting the stability of foams

1. drain gas bubbles together with liquid films,
2. have gas bubbles with interfacial film ruptures,
3. cause diffusion of small to large gas bubbles, and
4. have a continuous phase in evaporation.

1.1.17 NUCLEATION

Some foods that contain crystalline materials are made up of colloids; however, not all colloids in foods contain this crystalline material. Examples of some foods, in which crystalline nucleation plays a fundamental role in food colloids during processing, are margarine, cream, fat spreads, milk fats, and confectionery creams (Povey, 21014).

If conditions are not far from equilibrium, then the formation of a new phase is often very slow. For a new phase to be large enough to grow spontaneously, nucleation, that is, the formation of small regions of the new phase in which it develops, must occur. In this chapter, the nucleation will be dealt with in a very brief way (Povey, 21014).

1.1.17.1 CRYSTALLIZATION OF COLLOIDS IN FOODS

Colloidal systems are used in the crystallization processes and participate in different ways: for example, as control agents or structure-directing agents, as templates for crystallization, and as scaffolds and nanoreactors, on the one hand. However, it is desirable to obtain particles in a crystalline or nanocrystalline colloid state (Muñoz-Espí, 2013).

In the nucleation kinetics, the transformation of the crystals melts a liquid in bulk that is dispersed as small droplets in such a way that it forms a colloidal substance, which includes the dispersion of particles and a phase of liquid suspension; these in turn are liquid and then are transformed into solids. The transformation of the liquid phase into a crystalline solid is partial, a transformation that occurs in many food colloids, in such a way that it gives rise to a dispersion of the solid within a dispersed liquid phase (Muñoz-Espí, 2013).

A very clear example of a colloidal system of this type is cow's milk; in the oil droplets that are dispersed in the aqueous suspension phase, particles of crystalline fat are present. What happens in this case is that the crystalline material depends on the feeding regime of the animals and time of year, which complicates things to obtain crystalline materials (Povey, 2016).

The first stages in the manufacturing of butter include nucleation of the particles of fat in oil droplets that make up the dispersion of milk (Povey, 2016).

1.1.17.2 NUCLEATION THEORY IN FOOD COLLOIDS

Currently, the applicability of the nucleation theory has been questioned because no quantitative predictions of the nucleation rate can be given. However, the trends and basic principles are illuminating and can be well predicted. Some authors explain the theory of nucleation in colloidal systems albeit in a modified form, on a theoretical basis in colloidal and bulk fluids. In particular, there is the idea that there is a critical nucleus size in which crystal growth begins and that at a certain size it is determined by an energetic balance between the volume and surface (Povey, 2014).

There are two types of nucleation: those of homogeneous volume and those of catastrophic nucleation; whatever the type is, the rate of isothermal crystallization can be modeled at an independent rate of nucleation because the diameter of the particle is assumed in the composition of the drop (Povey, 2014).

The result of a balance between a ΔG_S positive energy term and a ΔG_V negative volume term is the $\Delta G_{\text{nucleus}}$ energy change,in which the positive energy term ΔG_S is due to surface tension and the negative volume term ΔG_V is due to the enthalpy of fusion. Assuming that the nucleus is a sphere, the nucleation is isotropic, where g_i is a surface energy A_i is surface of each face, V is the volume of the nucleus, and ΔG_V is the change in Gibbs energy per unit volume arising from the phase transition (Povey, 2013, 2014).

$$\Delta G_{\text{nucleus}} = \Delta G_S + \Delta G_V = \sum A_i \gamma_i + V \Delta G_V \quad (1.17)$$

Then, the equation becomes

$$\Delta G_{\text{nucleus}} = 4\pi r^2 \gamma \mid 4/3\pi r^3 \Delta G_v \quad (1.18)$$

1.1.18 SOFT SOLIDS

The consistency of food plays a determining factor in the way by which the user experiences the consumption of such products. Soft solids provide a widely popularized alternative found in a wide range of foods such as bread, cheese, meat, and pudding. The term soft solids can be defined by considering both words individually and in correlation. A solid is a material that when exerting a tension on it will produce mainly elastic deformation. On the other hand, the word "soft" determines that the tension for such deformation is relatively small (Zúñiga and Aguilera, 2008).

Soft solids can be defined as fluids that do not adhere to the same rules as normal liquids, that is, they are non-Newtonian fluids. Fluids such as cooking oil or water do not change their behavior because they are Newtonian fluids. An example for this is the mixing of water for 1 h at a high speed in a cup—water will begin to flow the same way at the end of 1 h as at the beginning (Zúñiga and Aguilera, 2008).

However, the mechanisms that make soft solids distinctive in this way are complex and have not been well understood yet, and this makes controlling their properties difficult (Lazidis et al., 2017).

1.1.18.1 GELS

A gel is a three-dimensional polymer network containing a solvent, which is distinguished by mechanical rigidity during the observation time but with almost the same density of the solvent. The gels contain two phases:

a continuous or solid phase, which is characterized by having a long range of particles, a structure at the molecular level, and the dispersed or liquid phase, so both phases form a water/solid complex at room temperature (Van Doorn et al., 2017).

The ability of food polymers to form gels depends on formation of areas of union between the polymer molecules that restrict the expansion of the network. The liquid is therefore entrapped by polymer molecules that are aggregated into one immense molecule with a three-dimensional structure (Van Doorn et al., 2018).

The formation of the structures will depend on several factors such as the gradual production in situ or the external controllable conditions such as temperature, pressure, gelling time, or the composition of the solution (ionic resistance, pH, solutes of low molecular weight, among others) (Koç et al., 2014; Zúñiga and Aguilera, 2008).

The rheological and textural properties of the gels are conferred to the network structure and its relationship with the liquid phase. Therefore, gels cover a wide spectrum of textures, as illustrated in Figure 1.5.

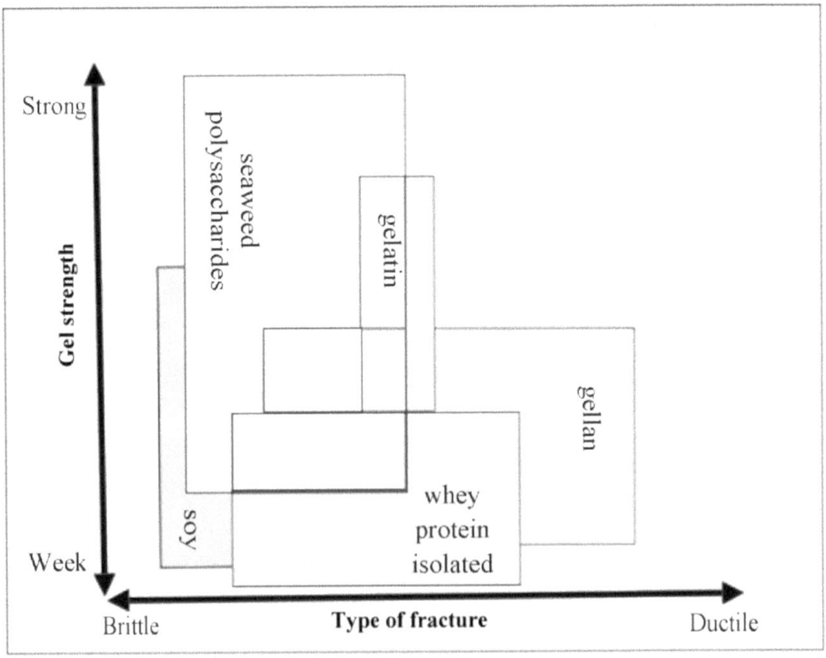

FIGURE 1.5 Edible gels: resistance and fracture properties.

1.1.19 RHEOLOGY OF SOLIDS

The study of attributes and mechanical properties is essential for soft solids, as they affect their consistency during consumption and handling, as well as the physical stability of these solids. Soft solids usually have elastic modulus, so a module can be defined as the relationship between relative deformation and stress (Walstra, 2003). There are several types of modulus, as follows: G = shear modulus; γ = shear strain; E_u = Young's modulus or uniaxial elongational modulus; ε_u = uniaxial elongated tension; E_b = biaxial elongation modulus; and ε_b = biaxial elongation strain.

All moduli are related as shown below.

$$E_u = 2G\,(1 + m) \qquad (1.19)$$

$$E_b = 4G(1 + m)$$

Subsequently, the relationship of Poisson's ratio μ is given by

$$\mu = \frac{1}{2}(1 - \frac{d\ln v}{d\varepsilon}) \qquad (1.20)$$

In this equation, the second term in parentheses is the measurement of the change in volume n when a tensile stress is applied to the material. In most solid foods, the relative volume change is quite small, which is why the Poisson ratio is close to 0.5 and leads us to have $E_u = 3G$. Foods such as bread or spongy foods can have a value close to zero (Walstra, 2003).

1.1.20 QUANTUM CHEMISTRY

1.1.20.1 PRINCIPLES OF THE ATOMIC STRUCTURE

Atoms are made up of three subatomic particles: protons and neutrons make a massive nucleus and electrons "orbit" the nucleus in equal quantity with respect to protons. Protons have a positive charge, while electrons have a negative charge but with equal magnitude. However, neutrons do not have charge that are normally in equal numbers with respect to protons. In this context, *chemical elements* are defined by the number of protons (atomic number, Z). The variants of atoms with a different number of neutrons but same number of protons are named *isotopes* (Wade Jr., 2010).

The chemical properties of an element are determined by the quantity of protons and electrons. The electrons can form bonds and determine the molecular structure. The electrons are in *orbitals* around the nucleus. The *Heisenberg uncertainty principle* establishes that the position of an electron cannot be precisely determined. However, a probability function of the electron position in the orbital can be determined, which is the *electron density*. So, an orbital is an allowed spatial arrangement of energy of an electron, associated with a probability function that defines the electron density. The atomic orbitals are grouped in various layers around the atomic nucleus at different distances. In organic compounds, most of the common elements are in the first two rows (periods) of the *periodic table of the chemical elements*. The elements of the first period (defined by the quantum number $n = 1$) can contain up to two electrons, whereas the elements of the second period ($n = 2$) can contain up to eight electrons. Figure 1.6 shows the orbital structures and electronic configurations of the elements in the two main periods (Wade Jr., 2010).

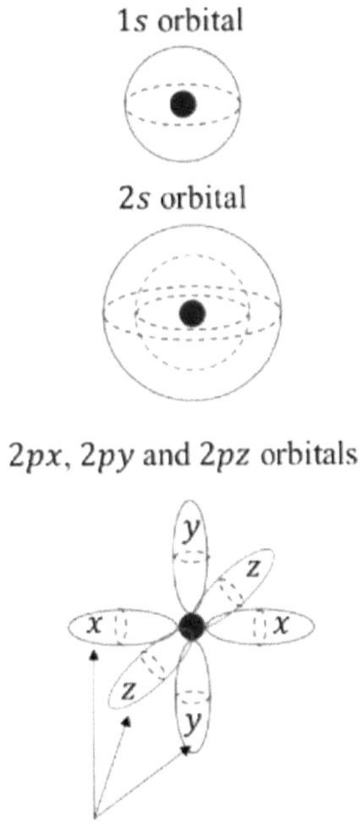

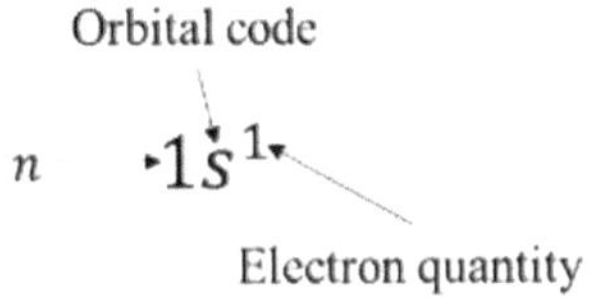

Element	Electronic configuration					Z
H	$1s^1$					1
He	$1s^2$					2
Li	$1s^2$	$2s^1$				3
Be	$1s^2$	$2s^2$				4
B	$1s^2$	$2s^2$	$2p_x^1$			5
C	$1s^2$	$2s^2$	$2p_x^1$	$2p_y^1$		6
N	$1s^2$	$2s^2$	$2p_x^1$	$2p_y^1$	$2p_z^1$	7
O	$1s^2$	$2s^2$	$2p_x^2$	$2p_y^1$	$2p_z^1$	8
F	$1s^2$	$2s^2$	$2p_x^2$	$2p_y^2$	$2p_z^1$	9
Ne	$1s^2$	$2s^2$	$2p_x^2$	$2p_y^2$	$2p_z^2$	10

FIGURE 1.6 Orbital structures and electronic configurations.

The *Pauli exclusion principle* says that each orbital could contain a maximum of two electrons. As a consequence, the layer $n = 1$ could contain up to two electrons and $n = 2$ up to eight electrons. The electrons of the external orbital are capable of forming chemical bonds, only if they are not *paired*. Naturally, the electrons repeal each other, and to make them fit, additional energy is required. The *Hund rule* establishes that when there are two or more orbitals with the same energy, electrons will be accommodated in separate orbitals. For instance, for $n = 2$, the electrons are distributed in an orbital *s* and two *p* orbitals. The orbital *s* is full, whereas the electrons in p_x and p_y are unpaired and p_z is empty (Wade Jr., 2010).

1.1.20.2 CHEMICAL BONDS

A theory that describe how atoms combine to form molecules is the *octet rule*. This says that a *full electron layer* is stable, and the atoms transfer or share electrons to achieve a full electron layer. In this sense, the electronic configuration of *noble gases* (He, Ne, Ar, Kr, Xe, and Rn) is considered a full electron layer. The elements with lower *Z* tend to achieve the electronic configuration of its reference noble gas. For example, carbon (with $Z = 6$) want to fill its last layer with eight electrons to achieve the electronic configuration of neon (with $Z = 10$). Therefore, carbon can form four chemical bonds (Wade Jr., 2010).

Figure 1.7 schematizes the three forms in which atoms could interact to achieve the noble gase configuration: covalent, ionic, and metallic bonds. In covalent bonding, the unpaired electrons of two atoms are paired to form one or more bonds. Covalent bonds occur between atoms of nonmetallic elements to form *volatile compounds*. In ionic bonding, pairing occurs between metallic atoms and nonmetallic atoms. The metallic atom transfers its unpaired electrons to the nonmetallic atom to form salts. Finally, the metallic bonds occur only between metallic elements. In this case, the metallic atoms with $n = 3$ or more layers can receive a considerable quantity of electrons to have a full layer and form great atomic lattices. However, most of these electrons are weakly attracted by the atomic nucleus, meaning that the lattice structure is bonded by a *sea of electrons*. In this context, in organic compounds (molecules mainly based on carbon, hydrogen, nitrogen, and oxygen), the most common chemical bond is a a covalent bond (Wade Jr., 2010).

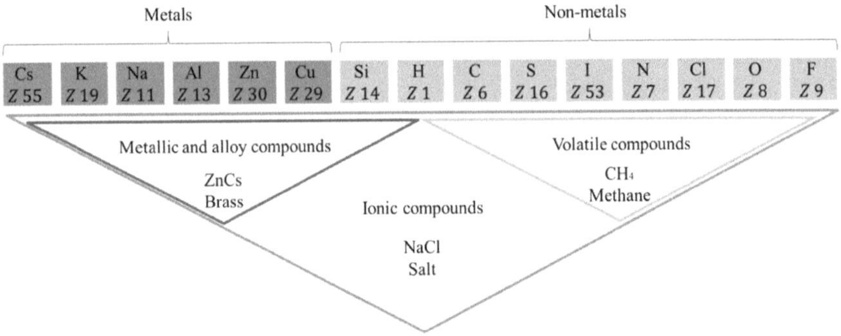

FIGURE 1.7 Three basic chemical bonds.

1.1.20.3 MOLECULAR ORBITALS

Atomic orbitals can be overlapped to generate more complex orbitals. To simplify the description of molecular orbitals, there is center only in molecular bonds and no *anti-bonds* which are normally empty in stable molecules. The stability of covalent bonds is caused by their high electronic density in the bonding region (the space between both atomic nucleus). Also, if both nuclei are too far from each other, the electronic interaction is reduced. On the other hand, if the nuclei are too near to each other, the electrostatic repulsion drives them away. In this context, the intermolecular distance at which the attraction and repulsion are in balance is named *bond length* (Wade Jr., 2010).

In organic compounds, the most common bonds are named sigma (σ) and pi (π). All simple bonds in organic compounds are sigma, whereas double or triple bonds contain sigma and pi orbitals. Figure 1.8 exhibits the origin and structure of sigma and pi orbitals in simple and complex bonds. A sigma bond is a product of the overlapping between two *s* orbitals, an *s* orbital and an axial *p* orbital, or two axial *p* orbitals. On the other hand, pi bonds result from the combination of two perpendicular *p* orbitals. With respect to complex bonds, a double bond requires four electrons in the bonding region. The first pair is a sigma bond and the second pair is a pi bond. In turn, triple bonds are formed by two pi bonds and one sigma bond (Wade Jr., 2010).

Using *s* and *p* orbitals, the angles of organic molecule bonds can be predicted at 90°. However, the observation show that these angles generally are 109°, 120°, and 180°. A common explanation of this observation is the *valence shell electron pair repulsion* theory. Around a central atom, because the electron pairs repeal each other, the bonded and nonbonded electron pairs are separated by the largest angles. To explain the common organic molecular shape, it is assumed that the *s* and *p* orbitals are combined to form *hybrid atomic orbitals*. Figure 1.9 schematizes the common hybrid atomic orbitals in organic compounds. A bond linear arrangement (180°) is a result of an *s* orbital and a *p* orbital, named as *sp* orbital. Meanwhile, 120° bond angles are required to orient three bonds in such a way that they are as far as possible. This configuration is the result of a hybridization between an *s* orbital and two *p* orbitals and is named sp^2 orbital. Finally, four hybridized bonds require 109.5° angles and are a result of the hybridization of an *s* orbital and all of the *p* orbitals. These hybridized bonds are present in a lot of organic compounds in foods. For instance, vitamin C (present in a lot of edible fresh fruits and vegetables) have sp^2 and sp^3 bonds. On the other hand, to explain the macroscopical physical chemistry changes observed in foods, understanding of the interaction between molecules is also necessary (Wade Jr., 2010).

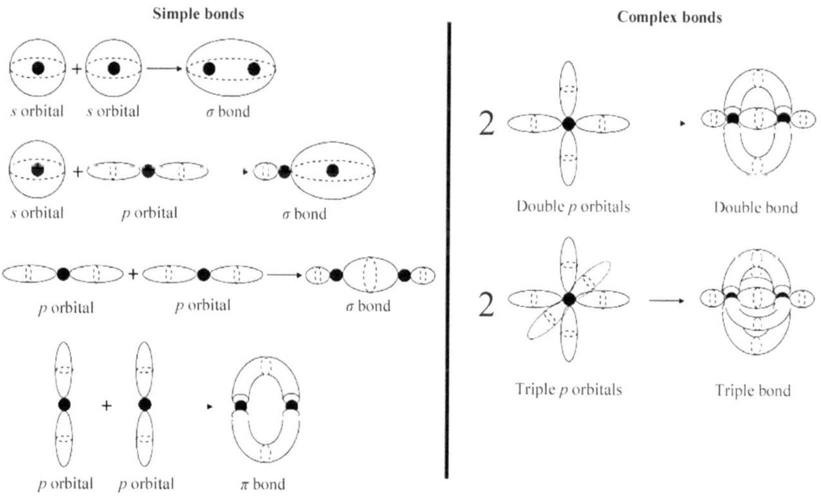

FIGURE 1.8 Common molecular bonds in organic compounds.

1.1.20.4 POLARITY AND MOLECULAR INTERACTIONS

In general, when a molecule is formed, the electrons in the bond could be equally attracted by both sides of the molecule; this is named a *nonpolar*

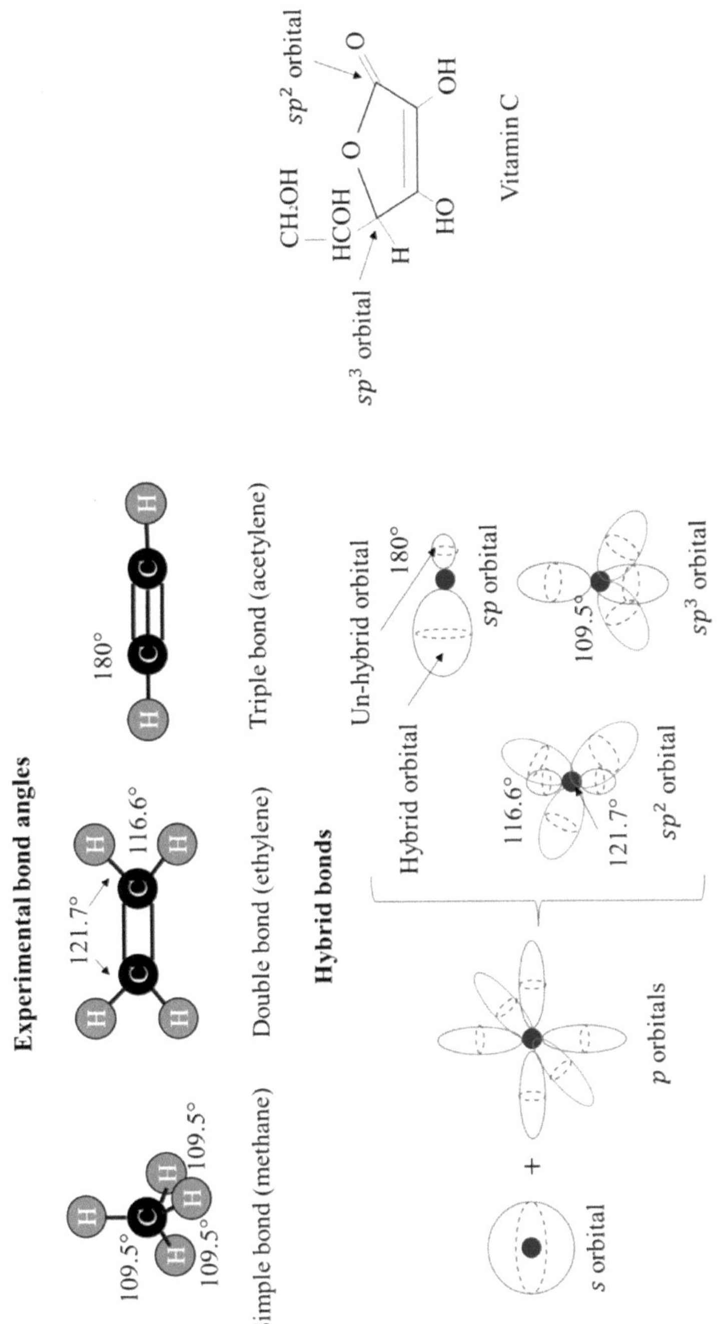

FIGURE 1.9 Hybridized orbitals common in organic molecules.

covalent bond. On the other hand, if one of the bonded atoms attract strongly the shared electron pair, it is known as a *polar covalent bond.* Pauling electronegativity (Figure 1.10) is used by organic chemists to predict the polarity of covalent bonds. Generally, elements with higher electronegativities (attraction force exerted by the atomic nucleus on the electrons) attract strongly the bonded electrons. So, the atom with the highest electronegativity polarizes the bond. In this context, when a bond is polarized, the molecule presents a partial charge, that is negative (δ^-), on the more electronegative molecule side. In the δ^- region, there is higher electron distribution, whereas in the positive partial charge (δ^+) region, the probability distribution is lower. The bond polarity is measured by the *dipolar moment* (μ), which is the difference between δ^- and δ^+ multiplied by the bond length (Wade Jr., 2010).

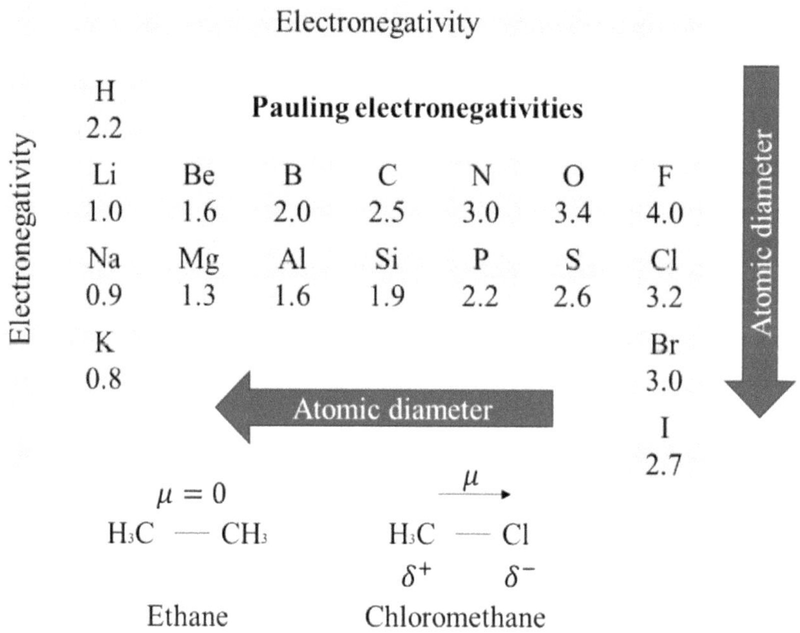

FIGURE 1.10 Pauling electronegativity.

However, when two molecules approach each other, they either attract or repeal. These interaction forces are known as *van der Waals forces.* Interaction forces between molecules are relevant in solids and liquids because their molecules are continuously in contact. Physical and chemical properties (like solubility, fusion, and boiling point) depend strongly on the van der Waals forces. There are three main kinds of attraction forces in organic molecules: London dispersion forces, dipole–dipole forces, and hydrogen bonds. These forces are schematized in Figure 1.11 (Wade Jr., 2010).

In nonpolar molecules, the main attraction force is named *London dispersion*, and it emerges from temporal dipoles induced by the nearest molecules. Contrary to this, polar molecules have permanent dipoles, as result of their polar bonds, that produce stronger attraction forces between molecules, known as *dipole–dipole force*. Finally, the strongest dipole–dipole attraction force is called *hydrogen bond* (which is not a real bond). Particularly, in organic compounds, hydrogen bonds occur only on hydrogen atoms chemically bonded to nitrogen or oxygen atoms (Wade Jr., 2010; Walstra, 2003).

The van der Waals forces are especially relevant to explain the microscopical properties of thermodynamic systems like surface phenomena, dispersion and the formation of solutions, emulsions, foams, and colloids. On the other hand, the next section describes the study of thermodynamic systems on a macroscopical scale (without involving molecular interactions).

1.1.21 REACTION KINETICS

The rate and equilibrium of a chemical reaction mainly depend on the temperature, pressure, and concentration of reactants and products, symbolized as [A], where A is a chemical formula. Although the reaction rates are not susceptible to thermodynamic study, the reaction equilibria are. Despite many chemical reactions not reaching the equilibrium, taking in count the reaction rate and other considerations, such as heat and mass transfer, is crucial to make predictions. However, the equilibrium state provides reference to improve the chemical processes and allows us to determine whether the investigation of a new process is worthwhile. Before exploring the stoichiometry and the reaction equilibrium, it is necessary to know the basic concept such as the reaction coordinate (ε) and mole fraction (y) (Smith et al., 2017).

A general chemical reaction is written as $\sum_{i=\text{reactant}} |v_i| A_i \rightarrow \sum_{i=\text{product}} |v_i| A_i$, where v_i is a stoichiometric coefficient and A_i represents the chemical

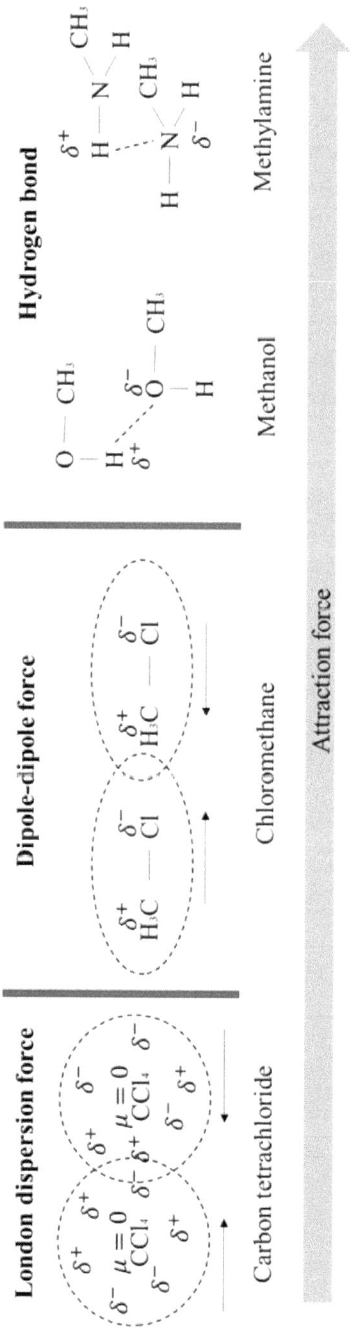

FIGURE 1.11 van der Waals forces.

formula. According to the sign convention for v_i, it is positive for a chemical product and negative for a reactant. For instance, the stoichiometric numbers for the reaction $CH_4 + H_2O \rightarrow CO + 3H_2$ are respectively $v_{CH_4} = -1$, $v_{H_2O} = -1$, $v_{CO} = 1$, and $v_{H_2} = 1$. On the other hand, as the reaction progresses, the changes in the number of moles (n) of chemical species (i) involved in the reaction are directly proportional to v_i. This relation is represented as a differential equation equalizing each other ($dn_i / v_i = dn_{i+1} / v_{i+1} = \cdots$) because the reagent consumption to be transformed in products is proportional. These relations can be identified collectively by a single quantity representing an amount of reaction symbolized as dε. This new variable (ε) is called the reaction coordinate and describes the degree of reaction. The reaction coordinate is useful to know the gas-phase molar fraction (y_i), which is a measure of each chemical species distribution. Table 1.2 comprises the formulas that conform the basic stoichiometry theory, where the subindex 0 indicates the initial unreacted state. These formulas correspond to a multireaction system, where each reaction is represented as j and the formulas emerge from specific mathematical identities and algebraic and integration procedures (Smith et al., 2017).

TABLE 1.2 Basics of the Stoichiometry Theory

$$n_i = n_{i0} + \sum_j v_{ij} \varepsilon_j \tag{1.21}$$

$$n_0 = \sum_i n_{i0} \tag{1.22}$$

$$n = n_0 + \sum_j \sum_i v_{ij} \varepsilon_j \tag{1.23}$$

$$v_j = \sum_i v_{ij} \tag{1.24}$$

$$y_i = \frac{n_{i0} \sum_j v_{ij} \varepsilon_j}{n_0 + \sum_j v_j \varepsilon_j} \tag{1.25}$$

Moreover, because ε is a single variable that characterizes the progress of the chemical reaction, and therefore the chemical composition of the system, the total Gibbs energy at T and P is determined by ε. The reaction coordinate has its equilibrium value (ε_e) when the Gibbs energy achieves its minimum value. Although the equilibrium expressions are developed for closed systems at constant T and P, they are not restricted in application to systems that are closed and thermodynamically balanced (Smith et al, 2017).

In this context, in open systems, the material may pass into and out of a system. As a consequence, the mathematical product $nG = f(n_i)$ and the Gibbs energy (described by Equation 1.17) need to be rewritten as Equation 1.21, where the partial differential relation term is defined as chemical potential (μ) of the species in the mixture (Smith et al, 2017).

$$d(nG) = (nV)dP - (nS)\,dT + \sum_i \left.\frac{\sigma(nG)}{\sigma(n_i)}\right|_{T,P,n}$$

$$dn_i = (nV)\,dP - (nS)dT + \textstyle\sum_i \mu_i dn_i \quad (1.26)$$

To directly relate nG with ε, Equation 1.21 must be used to replace n_i in Equation 1.26, resulting in a new expression described by Equation 1.27. This is understood considering that the initial unreacted state has no present chemical potential. Thus, the quantity $\sum_i v_i \mu_i$ represents the rate of change of nG with respect to the reaction coordinate at constant T and P and is zero at the equilibrium state (Equation 1.27) (Smith et al., 2017).

$$d(nG) = (nV)dP - (nS)dT + \sum_i v_i \mu_i d\varepsilon \quad (1.27)$$

$$\sum_i v_i \mu_i = \left.\frac{\sigma(nG)}{\sigma(\varepsilon)}\right|_{T,P} = 0 \quad (1.28)$$

The Gibbs energy is also related to other thermodynamic properties like the reaction equilibrium constant (K) described in Equation 29 and the standard heath reaction in Equation 1.30, where $\Delta M^\circ = \sum_i v_i M_i^\circ$, M_i° is the standard property (H, G, and Cp) of the chemical species and $Cp°$ is the standard heath capacity (Smith et al., 2017).

$$K = e^{\frac{-\Delta G^\circ}{RT}} \quad (1.29)$$

$$\frac{\Delta G^\circ}{RT} = \frac{\Delta G_0^\circ - \Delta H_0^\circ}{RT_0} + \frac{\Delta H_0^\circ}{RT} + \frac{1}{T}\int_{T_0}^{T} \frac{\Delta Cp^\circ}{R} dT - \int_{T_0}^{T} \frac{\Delta Cp^\circ}{R} \frac{dT}{T} \quad (1.30)$$

When the equilibrium state in a reacting system depends on two or more chemical reactions, the equilibrium chemical composition can be found by an extension of the expression developed for single reactions. Equation 1.31 is the basal description of the chemical species concentration for homogeneous and gas phase systems, which could serve as an introduction for more complex situations (Smith et al, 2017).

$$\prod_i y_i^{v_{ij}} = \left(\frac{P}{P^\circ}\right)^{-v_j} K_j \tag{1.31}$$

Deducing the concentration of the chemical species involved in a reaction is an important stage for the calculation of the rate of reaction (r_i). In this context, the limiting reactant is usually chosen as the basis for calculation. Therefore, the rate of disappearance of A_i, denoted by r_{A_i}, depends on the temperature and concentration. For many irreversible reactions, it can be written as the product of a reaction rate constant (k) and a function of concentrations, as shown in Equation 1.32 (Fogler, 2016).

$$-r_{A_i} = kf([A_i], [A_{i+1}], \ldots) \tag{1.32}$$

The dependence of $-r_{A_i}$ on the concentration of the chemical species involved is almost determined by experimental observation. However, the functional dependence on concentration may be postulated from theory, and experiments are required to confirm the proposal. One of the most common general forms of this dependence is the *power law model* (Equation 1.33). Here, the rate of reaction is the product of the individual reacting species (Fogler, 2016).

$$-r_A = kC_{A_i}^{\alpha} C_{A_{i+1}}^{\beta} \cdots \tag{1.33}$$

The exponents of the concentration in Equation 1.33 lead to the concept of *reaction order*. The order of a reaction refers to the powers to which the concentrations are raised in a kinetic rate law. The global reaction order g is the sum of the reaction order of each chemical specie. Generally, g can be deduced from the units of k. For example, the rate laws corresponding to a zero-, first-, second-, and third-order reaction are described by Equations 1.34–1.37, respectively (Fogler 2016).

$$-r_A = k \quad k \equiv \text{mol s/dm}^3 \tag{1.34}$$

$$-r_A = k[A] \quad k \equiv \text{s}^{-1} \tag{1.35}$$

$$-r_A = k\,[A]^2 \; k \equiv \text{dm}^3\ \text{s/mol} \quad (1.36)$$

$$-r_A = k\,[A]^3 \; k \equiv \text{dm}^6\ \text{s/mol}^2 \quad (1.37)$$

1.1.22 ELECTROCHEMISTRY

Ionizable substances, like salts, acids, bases and polyelectrolytes, partially dissociate into ions when they are dissolved in water. In this context, ionic species generally are reactive because of their electric charge, and the charge generally is shielded to a certain extent by the presence of ions with opposite charge, named counterions. This implies that the activity coefficient may be greatly diminished when the concentration of counterions is high (Walstra, 2003).

The ion activity coefficient ($\gamma_\pm$) depends on a large number of factors, but for low-ionic-strength solutions, electric shielding is by far the main factor. On this basis, the Debye–Hückel "limiting law" has been derived. An ion in solution is, on average, surrounded by more counterions than co-ions (the same charge), thereby to some extent shielding the charge of the ion. The attractive electric energy between ions of opposite charge tries to arrange the ions in a regular lattice. The attraction is stronger when the ions are on average closer to each other, and consequently, the higher the ion concentration, the stronger the shielding. The total ionic strength (I) is defined by Equation 1.38, where m denotes the molarity and z is the valency of ions (Walstra, 2003).

$$I = \frac{1}{2}\sum_i m_i z_i^2 \quad (1.38)$$

In the theory, the size of the ion also is involved and average $\gamma\pm$ of a dilute salt solution is regularly given by Equation 1.39, where ε is the relative dielectric constant. For instance, for water at 20°C, $\varepsilon = 80$, and this equation becomes Equation 1.40 (Walstra, 2003):

$$\gamma_\pm = \exp\left(\frac{42\times10^5}{(\varepsilon T)^{3/2}}\right)\left|z_+ z_-\right|\sqrt{I} \quad (1.39)$$

$$\gamma_\pm = \exp(-1.17)\left|z_+ z_-\right|\sqrt{I} \quad (1.40)$$

1.1.23 TRANSPORT PHENOMENA

A liquid food of not very high viscosity can be stirred to speed up the transport of heat or mass. Even if it contains dispersed particles, mostly are small enough to allow rapid diffusion in or out of them (see Table 1.3). However, many foods are solid like and there are even some that contain a lot of water. Cucumbers, for example, contain about 97% weight of water (Walstra, 2003).

TABLE 1.3 Diffusion Coefficients (D) of Some Molecules and Particles in Water at Room Temperature

Species	D (m²/s⁻¹	Diffusion Time in 1 cm
Water	1.7×10^{-9}	8 h
Sucrose	4.7×10^{-10}	30 h
Serum albumin	6.1×10^{-11}	10 days
Emulsion droplet (1 μm length)	4.2×10^{-13}	4 years

Generally, transport is by diffusion and in some cases by flow. For instance, when a food is kept, it may lose substances by diffusion or leaking, such as water or flavor components, or it may take up substances from the environment; or when a food is kept, the concentration of solutes, which may at first uneven, slowly becomes even. This may be of considerable importance for eating quality, likewise color substances may become evenly distributed. These processes often are slow, and transport rates may be difficult to predict. However, the study of idealized cases may be useful to understand the physical changes of food (Walstra, 2003).

Through a porous material, liquid may flow from the surface to the bulk, albeit often sluggishly. It is useful to consider the material as a solid matrix, containing several capillary channels or pores. In practice, the pores are always narrow enough and liquid velocity is slow enough to ensure that the flow is laminar. The superficial flow velocity (v), through a cylindrical capillary, is given by Darcy's law (Equation 1.41), where Q is the volume flow rate (m^3/s^{-1}), A is the cross-sectional area (perpendicular to the distance along the capillary, x) through which the liquid flows, B is the permeability (m^2), P is the pressure, and η is the viscocity. After this, the flow velocity on the bulk is described by Equation 1.41, known as the law of Hagen–Poiseuille, where r is the radius of the capillary (Walstra, 2003).

$$v = \frac{Q}{A} = -\frac{B}{\eta} = \frac{\Delta P}{\Delta x} \quad (1.41)$$

$$v = \frac{Q}{A} = -\frac{\Delta P}{\Delta x} \times \frac{r^2}{8\eta} \quad (1.42)$$

The permeability may be considered as a material constant, and it is a first approximation proportional to the square diameter of the pores in the material. However, in most real materials, the pore diameter shows a considerable spread. This implies that, by far, most of the liquids will pass through the widest pores. Moreover, the pores tend to be irregular in shape and cross section and are tortuous and bifurcate. Accordingly, the permeability may be anisotropic, and, as a consequence, it is not easy to predict B from the structure of the material. Taking this into account, numerous relations have been proposed, and one of the most used is the Kozeńy–Carman equation (Equation 1.42) (Walstra, 2003):

$$B \approx \frac{(1-\varphi)^3}{5A_{sp}^2} = \frac{(1-\varphi)^3 d^2}{180\varphi^2} \quad (1.43)$$

where φ is the volume fraction of the matrix material, A_{sp} is the specific surface are of the matrix (m^2 per m^3 of the whole material) and d is the diameter of an spherical material (the respective equation part is only valid for spherical materials of equal d). This equation is only valid for powders, so it has quite limited validity. However, it is useful to show that the permeability strongly decreases when the volume fraction and specific surface area increase (or the void volume decreases). For example, some very approximate magnitudes of B are 10^{-8} m^2 for ground coffee, 10^{-12} m^2 for curdled milk gel, and 10^{-17} m^2 for the polysaccharide gel. The very low permeability of several gels is the main reason why they hold water tenaciously. On the other hand, for very narrow pores (say 10 nm), the permeability depends on molecular size (Walstra, 2003).

1.1.24 BEYOND PHYSICAL CHEMISTRY

1.1.24.1 SURFACE PHENOMENA

As we have seen, most foods are dispersed systems and many structural elements constitute separate phases. This means that there are phase boundaries or interfaces, and the presence of such interfaces has several important consequences. Substances can absorb onto the interfaces, and when the interfacial area is large, the absorbed amounts can be considerable. The adsorption can strongly affect colloidal interaction forces between structural elements. Altogether, surface phenomena are of considerable importance during processing and for the physical properties of foods (Walstra, 2003).

In this context, fluid systems consisting of two phases try to minimize their interfacial area (like a blob assuming a spherical shape, the smallest surface area for a given volume). This is also commonly observed for large oil drops in water. When any system tries to minimize its free energy, it means that at an interphase the free energy is accumulated. This is called *surface of interfacial free energy*, and for a homogeneous interphase, the free energy must be proportional to the interphase area. Consequently, the surface is characterized by its specific surface free energy, which can be expressed in J/m^{-2} (Walstra, 2003).

The interphase formed in fluid systems will present a tension on its surface when a deformation force is applied, to minimize its interfacial area. This strength is named surface tension (γ), and it is expressed in units of N/m^{-1}. Table 1.4 presents some values of γ between solids and liquids (Walstra, 2003).

TABLE 1.4 Some Values of Surface Tensions Between Solids and Liquids.

Solid	Liquid	γ (N/m^{-1})
Ice (0 °C)	Water	25
Sucrose	Saturated sucrose solution	5
Triacylglycerol crystal	Triacylglycerol oil	4
	Water	31

On the contrary, the presence of solutes may affect the surface tension. For example, a solute can be preferentially accumulated on the surface, and after a fully packed monolayer has been obtained, no further accumulation takes place. Solutes showing this kind of behavior are called *surfactants*. On the other hand, the accumulation of a compound at an interphase is called *adsorption*. This is a very common phenomenon and can occur in all solid or liquids in contact with a gaseous or a liquid phase. The compound adsorbing is called the *adsorbate*, and the material onto which it adsorbs is the *adsorbent* (Walstra, 2003).

KEYWORDS

- **physical chemistry**
- **quantum chemistry**
- **multidisciplinary science**
- **thermodynamic Models**
- **food science**
- **engineering technology**

REFERENCES

Berk, Z. (2009). *Food process engineering and technology* (1st ed.). Amsterdam: Academic Press.

Chiralt, A. (2009). Thermodynamics in food engineering. In Gustavo V. Barbosa-Cánovas (Ed.) *Food engineering—Volume 1* (1st ed., pp. 150–176). Atlanta, USA:. EOLSS.

Green, A., Littlejohn, K., Hooley, P., and Cox, P. (2013). Formation and stability of food foams and aerated emulsions: Hydrophobins as novel functional ingredients. *Current Opinions in Colloid & Interface Science*, 18(4), 292–301. doi:10.1016/j.cocis.2013.04.008.

Fogler H. S. (2016). *Elements of chemical reaction engineering* (5th ed.). New Jersey: Pearson.

Holcroft, D. (2015). Water relations in harvested fresh produce. PhD thesis. The Postharvest Education Foundation (PEF). Retrieved from: http://www.postharvest.org/Water%20relations%20PEF%20white%20paper%20FINAL%20MAY%202015.pdf

Dincer, I., and Cengel, Y. A. (2001). Energy, entropy and exergy concepts and their roles in thermal engineering. *Entropy*, 3, 116–149. https://doi.org/10.3390/e3030116

Jones, R. T. (1997). Thermodynamics and its applications: An overview. Retrieved from https://www.coursehero.com/file/26357853/Thermodoc/

Kasaai, M. (2014). Use of water properties in food technology: A global view. *International Journal of Food Properties*, 17(5), 1034–1054. doi: 10.1080/10942912.2011.650339

Koç, H., Çakir, E., Vinyard, C. J., Essick, G., Daubert, C. R., Drake, M. A., and Foegeding, E. A. (2014). Adaptation of oral processing to the fracture properties of soft solids. *Journal of Texture Studies*, *45*(1), 47–61. https://doi.org/10.1111/jtxs.12051

Lazidis, A., de Almeida Parizotto, L., Spyropoulos, F., and Norton, I. T. (2017). Microstructural design of aerated food systems by soft-solid materials. *Food Hydrocolloids*, *73*, 110–119. https://doi.org/10.1016/j.foodhyd.2017.06.032

Levine, I. N. (2009). *Physical chemistry* (6th ed.): New York: Mc Graw-Hill.

Muñoz-Espí, R., Mastai, Y., Gross, S., and Landfester, K. (2013). Colloidal systems for crystallization processes from liquid phase. *Crystengcomm*, 15(12), 2175. doi: 10.1039/c3ce26657e

Povey, M. (2014). Crystal nucleation in food colloids. *Food Hydrocolloids*, 42, 118–129. doi: 10.1016/j.foodhyd.2014.01.016

Povey, M. (2016). Nucleation in food colloids. *The Journal of Chemical Physics*, 145(21), 211906. doi:10.1063/1.4959189

Roos, Y. (2003). Water activity: Principles and measurement. In *Encyclopedia of food sciences and nutrition* (2nd ed., pp. 6089–6094). Cambridge, Massachusetts: Academic Press–Elsevier. doi:10.1016/b0-12-227055x/01274-8

Ritzoulis, C., and Rhoades, J. (2013). *Introduction to the physical chemistry of foods*. New York: CRC Press.

Sandulachi, E. (2013). *Water activity concept and its role in food preservation*. Lecture, Technical University of Moldova.

Singh, R., and Heldman, D. (2009). *Introduction to food engineering* (4th ed.). Amsterdam: Elsevier/Academic Press.

Smith, J. M., Van Ness, H. C., Abbott, M. M., and Swihart, M. T. (2017). *Introduction to chemical engineering thermodynamics* (*8th ed.*). New York: Mc Graw-Hill

Toledo, R. (2007). *Fundamentals of food process engineering* (2nd ed.). New York: Springer.

Vaclavik V. A., and Christian E. W. (2008) Food emulsions and foams. In: *Essentials of food science*. Food Science Texts Series. New York, NY: Springer.

Van Doorn, J. M., Bronkhorst, J., Higler, R., Van De Laar, T., and Sprakel, J. (2017). Linking particle dynamics to local connectivity in colloidal gels. *Physical Review Letters*, *118*(18), 1–5. https://doi.org/10.1103/PhysRevLett.118.188001

Van Doorn, J. M., Verweij, J. E., Sprakel, J., and Van Der Gucht, J. (2018). Strand plasticity governs fatigue in colloidal gels. *Physical Review Letters*, *120*(20), 208005. https://doi.org/10.1103/PhysRevLett.120.208005

Walstra, P. (2003). *Physical chemistry of foods*. UK: CRC Press.

Welti-Chanes, J., and Vélez-Ruíz, J. (2003). *Transport phenomena in food processing* (1st ed.). Boca Raton, FL, USA: CRC Press.

Welti-Chanes, J., Mujica-Paz, H., Valdez-Fragoso, A., and Rios, L. (2009). Thermodynamics in food engineering. In Gustavo V. Barbosa-Cánovas (Ed.). *Food engineering—Volume I* (1st ed., pp. 240–274). Atlanta, USA:. EOLSS, Vol. 1.

Welti-Chanes, J., Vergara-Balderas, F., and Bermúdez-Aguirre, D. (2005). Transport phenomena in food engineering: Basic concepts and advances. *Journal of Food Engineering*, 67(1–2), 113–128. doi:10.1016/j.jfoodeng.2004.05.053

Zúñiga, R. N., and Aguilera, J. M. (2008). Aerated food gels: Fabrication and potential applications. *Trends in Food Science and Technology*, *19*(4), 176–187. https://doi.org/10.1016/j.tifs.2007.11.012

Wade Jr., L. G. (2010). *Organic chemistry* (7th ed.). New Jersey: Pearson.

Walstra, P. (2003). *Physical chemistry of foods*. New York: Marcel Dekker.

CHAPTER 2

PHYSICOCHEMICAL PROPERTIES AND EXTRACTION METHODOLOGIES OF AGROINDUSTRIAL WASTES TO PRODUCE BIOACTIVE COMPOUNDS

MARICELA ESMERALDA-GUZMÁN, CATALINA DE JESÚS HERNÁNDEZ-TORRES, DESIREE DÁVILA-MEDINA, THELMA MORALES-MARTÍNEZ, and LEONARDO SEPÚLVEDA*

Group of Bioprocess and Microbial Biochemistry, School of Chemistry, Autonomous University of Coahuila, Saltillo, Coahuila, México

**Corresponding author. E-mail: leonardo_sepulveda@uadec.edu.mx*

ABSTRACT

Currently, the food industry generates a large amount of agroindustrial waste that seriously affects the environment. This problem is spreading little by little around the world due to the lack of regulation and the inadequate treatment of these agroindustrial waste. The interest of modern biotechnology is to use agroindustrial waste to produce bioactive compounds that are of high added value and different uses in the industry. Bioactive compounds or secondary metabolites from plant origin, also called phytochemicals, are compounds with important biological properties and human health benefits such as antimicrobial, antiviral, antioxidant, anticancer, among others. This chapter will explain topics related to the physicochemical properties of agroindustrial wastes that are used in the north of the México to produce bioactive compounds. In addition, some characteristics of wastes of nutshell

(*Juglans regia*), grape (*Vitis vinífera*), and nopal (*Opuntia ficus-indica*) are described. Finally, some of the bioactive compounds that can be obtained by emerging technologies are used in the food industry.

2.1 INTRODUCTION

Disposal of waste has become a serious environmental issue, and the problem continues to increase with an increase in population and development of industries. An efficient and environmental solution is critically required to address the solid waste streams. The use of agrofood residues has gained interest from an economic and environmental point of view using unexploited biotic resources for metabolite production with increased value at low production costs; furthermore, by reducing environmental problems, it solves the problem of their disposal (Cholake et al., 2017; Schmidt, 2014).

Globally, significant amounts of agrofood residues are generated, most of which are burned as waste disposal. Given the abundance of this biomass, it can be used as a raw material to produce valuable chemicals through biochemical conversion. Research in this field has gained importance thanks to the concerns regarding the limitation of environmental problems and sustainability and preference toward natural, biodegradable, and environmentally friendly products. The use of an abundant and cheap source of natural compounds is in accordance with the concept of green chemistry, also known as environmentally benign chemistry or sustainable chemistry (Andrade et al., 2011; Diaz, 2018).

2.2 NUTSHELL (*JUGLANS REGIA*)

Nutshell (Figure 2.1) is a type of waste lignocellulose obtained from nut production processing and, at present, is subtilized and generally discarded or burned, which have caused serious resource waste and environmental pollution; these are the main routes to deal with this waste.

Disposal of nutshells has created serious problems for the processing industries, and hence, a sustainable recycling solution is critically required. Previously, the only solutions are to use as garden mulch, animal filler, and chicken litter; however, the conversion of this waste into a resource with increased value is an important pressing concern (Rajarao et al., 2014;

Fan et al., 2018). The most attractive consideration for nutshells use as an industrial product can be its low cost, abundant availability, and chemically reactive nature (Rodrigues et al., 2006).

FIGURE 2.1 Nutshell (*J. regia*).

The important and attractive properties of the shell waste are very low ash and high carbon content compared to other biomass such as coconut shell. It is therefore a potential carbon source for synthesizing materials, such as activated carbons (Poinern et al., 2011; Rodrigues et al., 2013). Nutshell liquid is a byproduct, cheap, and renewable and has many biological and industrial applications because it can easily react to form various derivatives, including polymers and resins. The nutshell shows biological activities such as antitumor, antioxidant, gastroprotective, and antibiotic properties. In the industry, the polymers and resins are widely employed as friction materials, surface coatings, adhesives, laminates, rubber compounding chemicals, flame retardants, and anticorrosive paints (Lubi et al., 2000).

The nut is rich in monounsaturated and polyunsaturated fatty acids, such as omega 3 and omega 6, is also a source of protein, rich in arginine, phytosterols, and phytochemical compounds, and also contains different vitamins and minerals, such as vitamin E, B complex, and iron. After consumption of the nut, there are different residues from which different compounds are obtained; for example, the Pecan nutshells, which represent between 40 and 50% of waste generated from the processing of nuts, are rich in phenolic compounds; a few phenolic compounds such as ellagic acid, gallic acid, chlorogenic acid, *p*-hydroxybenzoic acid, epigallocatechin, and epicatechin-gallate have been reported in pecan nutshell extract. Nutshells are an exceptional source of tannins and can be obtained with high yields (about 40% on a dry weight basis) and are more reactive than those from Mimosa, the industry standard for condensed tannins. However, the extraction of phenolic compounds from pecan nutshell may be difficult because they can bind to the matrix components such as cellulose and lignin. Furthermore, extraction conditions as the pH of the solution, temperature, time, solvent concentration, pressure, and powder nutshell particles may contribute to the efficiency of extraction (Hilbig et al., 2018; Pinheiro et al., 2013). From the mesocarp and pericarp of the walnut, an oil is extracted that is used as a base for the elaboration of some colorants. However, these two outer layers of the walnut, together with the leaves, contain abundant tannins that have high strongly astringent (Guerra-Olgin, 2012; Koch et al., 2014).

The components of the nutshell vary a lot with the extraction process, which can be by a heating process as a thermo-mechanic process, in the cold in solvents, or by pressing. The nutshell liquid is a natural source of saturated and unsaturated long-chain phenols (anacardic acids, cardanol, and cardols) and a mixture of meta-alkylphenols with variably unsaturated benzene rings, in different percentages depending on the extraction method, time, and temperature at which the nutshell was submitted (Lomonaco, 2009; Rodrigues et al., 2011). These compounds can be incorporated by erythrocytes and liposomal membranes, exerting antioxidant, antigenotoxic, and cytostatic activities. The ability of these compounds to inhibit bacterial, fungal, protozoan, and parasite growth seems to depend on their interaction with proteins and/or on their membrane-disturbing properties (Stasiuk and Kozubek, 2010; Andrade et al., 2011).

Anacardic acid, which is the major component of the nutshell liquid, has attracted great research interest due to its biological activities such as antitumor, antioxidant, gastroprotective, and antibiotic. In addition, it has been used as a synthon to produce a variety of biologically active compounds with increased efficiency, and some of them outperform their corresponding standard materials. Besides the biological activities, anacardic acid has recently been found to be a potential candidate as a capping agent for the development of nanomaterials (Hamad et al., 2015; Mlowe et al., 2014).

The proximate composition and mineral concentration of cashew nut (*Anarcadium occidentale*) were investigated using standard analytical methods. The physicochemical characteristics of the cashew nutshell liquid were also determined. The proximate composition (%) is as follows: moisture (7.2), ash (2.8), crude fat (49.1), crude protein (36.3), crude fiber (3.2), and carbohydrate (by difference) (1.4). The mineral composition (mg/100 g) of cashew nut showed potassium (27.5 ± 0.4), to be the highest, calcium (21.5 ± 0.0), magnesium (19.3 ±0.1), sodium (8.2 ± 0.2), and phosphorous (14.0 ± 0.2). Zinc and iron concentrations were low. The physicochemical properties of cashew nut oil are as follows: color (yellow), refractive index (1.458), specific gravity (0.962), acid value (10.7 mg KOH/g), saponification value (137 mg KOH/g), iodine value (41.3 mg iodine/100 g), and free fatty acid (5.4 mg KOH/g). This is an indication that the oil is nondrying, edible, and may not be used for soap making. The nutshell liquid extracted was dark brown. The ash and moisture contents (%) were 1.2 and 3.9 for BRZ species and 1.3 and 6.7 for AFR species, respectively. The specific gravity and refractive index were 0.941 and 1.693 for BRZ variety and 0.924 and 1.686 for AFR variety , respectively. The saponification value and acid, free fatty acid (mg KOH/g), and iodine (mg iodine/100 g) contents were 58.1, 12.1, 6.1, and 215 for BRZ species and 47.6, 15.4, 7.8, and 235 for AFR species, respectively (Akinhanmi et al., 2008).

2.2.1 EXTRACTION METHODOLOGIES TO OBTAIN BIOACTIVE COMPOUNDS FROM NUTSHELL

Extractions of components have been known decades ago since many civilizations used different kinds of methods like cold pressing, solvent extraction, among others. Extraction is a common processing method used in the food industry. Driven by technical, scientific, and economic impediments associated with traditional extraction techniques, such as high energy cost, residual solvent impurities, and thermal degradation, in the past decade, the food industry has experienced a revolution in the development of greener technologies for the recovery of active ingredients (Ekezie et al., 2017). Nowadays, most of the industries use different processes to extract the components; however, there is a search for different techniques that allowed industries to reduce the costs of production. Existing extraction technologies often requiring up to 50% of investments in a new plant and more than 70% of total process energy used in industries (Chemat et al., 2017). Ultrasound-assisted extraction (UAE) has gained attention due to its green impacts on bioactive compound extraction, higher product yields, shorter processing time, and low maintenance costs (Barba et al., 2016; Chemat et al., 2017; Wen et al., 2018). The conventional techniques, such as maceration, infusion, and "Soxhlet" extraction, are time-consuming and use large amounts of solvents. The solvents commonly used for the extraction are chloroform, carbon tetrachloride, tetrachloroethylene, and chlorobenzene (chlorinated solvents) and acetone, methanol, and acetonitrile (nonchlorinated solvents), but the use of these solvents depends on the properties of the matrices. Ultrasound is considered green for the environment and it considers minimizing or eliminating the use of organic solvents (Tiwari, 2015).

It is known that UAE has great productivity; it only takes minutes to do a full extraction, giving the final product in higher purity; it has a high reproducibility in different types of matrices like microalgae, fruit and vegetable waste, plants, among others. It has been reported that this technique reduces the consumption of solvents, simplifying manipulation and workup, eliminating post-treatment of wastewater, and consuming only a fraction of the energy normally needed by a conventional extraction method. UAE has been used for different compounds such as pigments, antioxidants, aromas, polyphenols, organic and mineral compounds, and many others (Vinatoru et al., 2017).

The main physical parameters in the ultrasound process are power, frequency, and amplitude. During the UAE, different effects are produced like vibration, crushing, mixing, thermal, and cavitation; these effects are responsible for breaking the cell wall, liberating the desired components without causing changes in the structure and function of the extract, making the process successful (Wen et al., 2018).

Cavitation is an important part of the ultrasound method, consisting of the presence of bubbles in liquids; as the ultrasound wave propagates longitudinally in the liquid, its alternating pressure is periodically stretched and compressed. Cavitation bubbles vary due to the continuous compression and the rarefaction cycle as they grow to reach their critical value. It has been reported that temperature, pressure, and volume affect the cell structure, causing the process of mass transfer. Another effect caused by cavitation is the change of chemical processes in the system by initiating new reaction mechanisms or improving the speed of the process. The formation of free radicals is common in this type of process, especially when water is used as a solvent. Due to the destruction of water molecules, highly reactive free radicals can be produced, which can modify proteins. On the other hand, when a microjet is generated due to the collapse of the cavitation bubbles, different processes will take place such as surface peeling, erosion, breakdown of cell walls, and the exudation of cellular contents so that various compounds can be extracted (Chemat et al., 2017; Khadhraoui et al., 2018; Wen et al., 2018).

Different authors have reported that the yield after the ultrasound extraction process was higher than the one using different solvents. Guandalini et al. (2019) evaluated the use of ultrasound to extract sequentially phenolics and pectin from the mango peel. They reported that the best total phenolics yield (67 %) was obtained with an extraction solution consisting of 50% of ethanol in water (v/v) and without ultrasound application; however, the residue of this extraction was then used to extract pectin assisted by ultrasound and obtained an increased pectin extractionyield of above 50% without affecting its quality.

Microwave-assisted extraction uses microwave radiation as the source of heating of the solvent–sample mixture and is useful in analytical procedures and for the optimization of extracts for products. To extract components with this technology, different solvents are applied; ethanol/water and methanol/water are particularly useful in the extraction of phenolic and flavonoid compounds. When the solvent used during the process is water, microwaves interact selectively with the free water molecules present in the glands and vascular system, leading to rapid heating and temperature increase, followed by the rupture of the walls and release of the components into the solvent (Lidia et al., 2017). During the process of microwave-assisted extraction, there are some factors to consider for obtaining the best results, such as the nature of the material, solvents, temperature, time, solid–liquid ratio, pressure, particle size, and other parameters that could affect the extraction process. The need for modified microwave equipment to obtain better yield and to reduce the use of solvents or to avoid them completely has been studied. Researchers have reported the extraction of terpenoids, phenolics, alkaloids, polysaccharides, steroid saponins, and essential oils using microwaves and obtained better results than the traditional techniques (Ameer et al., 2017; Chupin et al., 2015; Lidia et al., 2017).

Kaderides et al. (2019) reported that the microwave method is more efficient that ultrasound method when it is applied to pomegranate peel to obtain phenolics of interest; the results obtained showed that with the microwave extraction method, they had a yield about 1.7 times higher in a shorter process time (4 min) in comparison to the UAE (10 min). Maran et al. (2015) used microwave-assisted extraction to obtain pectin from waste mango peel; the optimum microwave-assisted extraction conditions for the highest pectin yield (28.86%) from waste mango peel are as follows: a microwave power of 413 W, a pH of 2.7, a time of 134 s, and a solid–liquid ratio of 1:18 g/mL. Pandit et al. (2015) also used mango peel to extract pectin with the microwave extraction method. The yield of pectin was found to be maximum from the mango peel exposed at a microwave energy of 1000 W for 20 min; pectin extracted at the optimum conditions has the galacturonic acid content, methoxyl content, and viscosity of 57.2 g/100 g, 8.2 g/100 g, and 98.2 mPa s, respectively, resulting in a higher yield of pectin. Also, Xu et al. (2016) report the use of the microwave method and UAE to obtain juglone from walnut green husk (*Juglans nigra*) with a yield of 836.45 μg/g under the optimum conditions of a ratio of solvent to sample of 309.70:1, an ultrasonic power of 585.42 W, an ultrasonic time of 25.57 min, and a microwave time of 103.27 s.

2.3 GRAPE (*VITIS VINÍFERA*)

Grape (Figure 2.2) in Mexico is one of the fruits with greater agricultural activity. In addition to its consumption in fresh, the industrial grape is the main input for the wine sector, which represented 22.93% of the total production of grapes in 2016. It is estimated that by 2024, the production reaches 415.43 thousand tons with an annual growth of more than 1.8% (SAGARPA, 2017). Although the production of grapes is very promising, the generation of waste from this fruit is a real problem. However, these wastes can be exploited because they contain a wide range of polyphenolic compounds and other molecules with biological activities that are beneficial for the health of human beings. In this section, we will describe the most important extraction methodologies to obtain bioactive compounds from the use of grape wastes.

FIGURE 2.2 Grape (*V. vinífera*).

2.3.1 EXTRACTION METHODOLOGIES TO OBTAIN BIOACTIVE COMPOUNDS FROM GRAPE WASTE

Twenty-nine samples of grape marc from Spain and Italy were analyzed by chromatography to determine their major volatile compounds. In this research, the results obtained showed that waste samples contained a significantly higher concentration of methanol, 2-butanol, ethyl acetate, and ethyl lactate. Most of the samples were obtained by a simple distillation method (Cortés et al., 2011). Further, solid–liquid extraction of the total phenolic content from grape marc was studied. For this method, ethanol was used as an extraction agent and kinetically investigated. The yields of extraction ranged from 11 to 22 mg GAE g, with values of the extraction rate between 0.040 and 0.1302 min at a temperature range of 25–60 °C. The authors mentioned that polyphenols are mainly in the samples and have activity antioxidant (Sant´Anna et al., 2012). In another research, extracts rich in monomeric anthocyanins and total phenolic compounds were obtained from grape marc by pressurized liquid extraction. The extraction method was performed using ethanol and water mixtures (acidified or not) (50% w/w), pure ethanol, and acidified water at temperatures from 40 °C to 100 °C. The best results for anthocyanins extraction (ethanol–water pH 2.0 [50% w/w]) resulted in 10.21 mg of malvidin-3-*O*-glucoside/g of dried grape marc. The authors concluded that this research must be carried out to concentrate and extract the target compounds of the samples and these molecules have high bioactivity and are beneficial for human health (Tamires et al., 2019).

Subcritical water extraction is a modern extraction technique that posits a few advantages over traditional solvent extraction. The subcritical water extraction uses water at elevated pressures and temperatures to extract polyphenols compounds mainly of different polarities, through adjusting the water polarity by changing the system temperature. The authors concluded that a technoeconomic analysis of the extraction of bioactive phenolic compounds from New Zealand grape marc gave the same yields, was the best, and had low cost versus solvent extraction techniques (Richard and Saeid, 2017). In another research, a method based on pressurized solvent extraction to determine main polyphenolic compounds in the grape marc obtained as a byproduct of the white winemaking process was developed. The authors developed a pressurized solvent extraction method that is more advantageous than those based on traditional techniques such as maceration/agitation. Gallic acid, catechin, and epicatechin were the main polyphenolic compounds; these compounds present antioxidant activity (Álvarez-Casas et al., 2014).

Microwave-assisted extraction enhances antioxidant extraction kinetics in grape marc processing. This study focused on the dielectric property measurement of grape marc using a resonant cavity measuring device. These authors concluded that the results will provide useful information about the design of a microwave cavity to maximize the treatment efficiency of grape marc in a microwave industrial extraction process (Sólyom et al., 2013). The extraction by ultrasound is the methodology mostly used at present to obtain polyphenols compounds. In another research, the effects of acoustic energy density (6.8–47.4 W/L) and temperature (20–50 °C) on the extraction yields of total phenolics and tartaric esters during UAE from grape marc were investigated. The authors concluded that ultrasound technology proved to be more effective than conventional technologies during extraction from grape marc through performance comparison (Yang et al., 2014). Using the same microwave-assisted extraction technology, the phenolic compounds from Chardonnay grape marc were studied. The results reported was the optimal parameters found by experimental design were 48% ethanol for the solvent content, 10 min for the extraction time, and 1.77 g for the solid mass. Furthermore, the polyphenols obtained by microwave-assisted extraction showed high antioxidant activity. The authors concluded that the potential use of this extract can be a bioactive additive in protein film-forming formulations for food and pharmaceutical applications (Garrido et al., 2019).

There are several methods for the extraction of polyphenolic compounds and other molecules. The research studies focus on the conditions of extraction and the comparison of yields. Grape wastes are attractive and underutilized due to the large number of bioactive compounds that can be obtained; this represents an alternative to produce a new bioactive that can be used in the pharmaceutical, cosmetic, and food industries, among others.

2.4 NOPAL (*OPUNTIA FICUS-INDICA*)

Prickly pear *O. ficus-indica* (Figure 2.3) is a perennial plant belonging to Cactaceae family; its Crassulacean acid metabolism allows it to grow in places with low water availability, such as arid and semiarid areas and the treetops or as hydrophytes in places with limited availability of CO_2.

The plant is native to arid regions of North America, mainly Mexico and southern United States (Pérez-Méndez et al., 2015); however, it was introduced to Africa by the Spaniards and spread to several arid regions, including Australia, the Mediterranean Basin, and in some regions of South America such as Argentina (Marin-Bustamante et al., 2018). It is known by different names: in Mexico, it is called nopal; in Spain, chumbera; in India, fig of the Indies; in Italy, Fico d'India; in France, Figure 2.2. de Barbarie (Morocco, Tunisia, Eritrea, and Ethiopia) (Salehi et al., 2019).

Although there are more than 300 species of the genus of *Opuntia*, very few have been exploited (Torres-Ponce et al., 2015). In Mexico, the nopal has been traditionally used as food and is closely related to culture. The plant consists of cladode, flower, a fruit called "tuna," and seeds. Traditionally, the thorns are removed from young cladodes and they are cut into small pieces (nopalitos) for consumption, in different traditional Mexican dishes (Pérez-Méndez et al., 2019). The fruit (tuna) formed by the pulp, peel, and seeds is also edible (Andreu et al., 2018); its other applications are fodder (cattle and sheep) and industrially for alcohol production, dyes (cochineal), soap, pectins, and oils (Torres-Ponce et al., 2015).

FIGURE 2.3 Prickly pear of *O. ficus-indica.*

Since ancient times, the nopal has been used as a natural disease-fighting medicine due to health-promoting substances present in the plant (FAO, 2018).

Opuntia species are distributed worldwide and have great economic potential; both cladodes and the fruit are rich in bioactive compounds, which vary depending on the species, cultivation, and climate conditions (FAO, 2018) (Eleojo et al., 2018).

Such diversity of bioactive compounds like phenolic, nonphenolic, and pigments has promoted and diversified the industrial applications of the nopal plant (Eleojo et al., 2018).

At present, scientific studies have demonstrated the presence of bioactive compounds in both cladodes and fruits, findings that the plant is rich in fiber, minerals, flavonoids, phenolics, among other nutrients, while the fruit pulp is rich in glucose, fructose, pectin, ascorbic acid, flavonoids, betalains, and phenols (Torres-Ponce et al., 2015).

2.4.1 BIOACTIVE COMPOUNDS IN THE DIFFERENT PLANT COMPONENTS OF THE NOPAL (O. FICUS-INDICA)

2.4.1.1 CLADODES

The cladodes of *Opuntia,* normally used as food, contain dietary fiber, vitamins, minerals, trace elements, and phytochemicals. Their chemical composition depends on the species and variety, the plant age, and growth stage, as well as environmental conditions (Pérez-Méndez et al., 2015). Among the active compounds extracted from cladodes are the following:

Mucilage: It is a polymer with a highly branched structure that contains residues of arabinose, galactose, galacturonic acid, rhamnose, and xylose. The most common industrial applications are as a food, a food emulsifier, a thickener, and an adhesive. Commonly, the extraction is performed by a solid–liquid process where cladodes are macerated in water at room temperature (Felkai-Haddache et al., 2016).

Pectin: It is a natural product with wide applications in the industry; among its applications are a thickening agent, gelling and binder, water stabilizer, fixer in jams and candies, and viscosifier in soft drinks. Pectin extraction is carried out by conventional methods, such as solid–liquid extraction; this operation must be rapid to prevent degradation as this operation must be rapid to prevent degradation because it can change its functionality (Lefsih et al., 2017).

Dietary fiber: It improves the stability of food during production and storage as well as reduces the risk of degenerative diseases such as obesity, diabetes, and heart diseases. A lot of studies show that young cladodes present high antioxidant properties, while mature (2 years) have a higher amount of fiber. Despite fewer industrial applications, *Opuntia* cladodes have great potential (Cheikh et al., 2018).

2.4.1.2 FRUIT (TUNA)

The fruit pulp is sweet and can be of various colors such as greenish-white, yellow, orange, red, or purple with a thin shell. Containing a variety of active compounds in addition to vitamin C, carotenoids, and dietary fiber, the pulp can also be used for the production of natural dyes (betalains) (FAO, 2018).

Betalains: These are classified into two groups depending on their structural characteristics and light absorption: betacyanins are responsible for red and purple and betaxanthins for yellow color. In addition, they also present a high antioxidant and anticarcinogenic activity and anti-inflammatory properties (Cheikh et al., 2018).

2.4.1.3 FLOWERS

The flowers of different plants have shown medicinal properties; in the case of flowers of *Opuntia*, Ammar et al. (2015) noted that the methanolic extracts of flowers *O. ficus-indica* exhibit antioxidant and antiulcer activities, and they assessed the extraction yield using various solvents and maceration and Soxhlet methods (García-Cayuela et al., 2019). They concluded that the method and the type of solvent directly affect both the extraction yields and the profile of the obtained compounds. The *Opuntia* flowers have many uses, such as a raw material for the production of bioactive compounds in the pharmaceutical field and as preservatives in food (Andreu et al., 2018).

2.4.1.4 RESIDUES

In Mexico, the process of removing the thorns generates 40,000 tons of waste, of which 90% is the skin of the cladodes that is rich in pectin, cellulose, and water and the other 10% corresponding to the thorns contains primarily cellulose and lignin.

2.4.1.5 THORNS

One application with potential for this residue is in the production of biodegradable nanomaterials as nanocellulose for its application as a reinforcement in edible films, food packaging, the structure of bioelectronic compounds, among others (Marin-Bustamante et al., 2018) (Marin-Bustamante et al., 2017).

2.4.1.6 FRUIT PEEL

The main components of the shell are cellulose, hemicellulose, pectin, proteins, antioxidants, mineral flavonoids, and other polysaccharides whose applications are mainly as food (Barba et al., 2017). In another research, the profile of bioactive compounds present in the peel of *O. ficus-indica* and *Opuntia engelmanni* was studied, highlighting their chemical composition, antimicrobial and antioxidant capacity of the extracts, and the lack of toxicity (Melgar et al., 2017). On the other hand, concluded that the peel extracts have high potential as nutraceuticals, however, suggested that the extraction methods should be more sensitive to identify more compounds (Eleojo-Aruwa et al., 2019).

2.4.1.7 SEEDS

Seeds contain oil rich in substances that are beneficial for health, such as unsaturated fatty acids, vitamin E, carotenes, and other bioactive compounds (Barba et al., 2017). However, due to their low performance, the oil is not attractive as edible oil, so its use has remained only in the pharmaceutical and cosmetic industries (FAO, 2018). The extraction is carried out by supercritical carbon oxide or by Soxhlet with petroleum ether, hexane, or chloroform:methanol at different ratios (Eleojo et al., 2018).

2.5 EXTRACTION METHODS

For bioactive compounds to be commercialized, the process has to be scaled from the laboratory to industry, and certainly, this is a major challenge that requires several stages, including the processes of extraction, purification, contaminant separation, toxicological studies, in vitro and in vivo assays, bioavailability, bioaccessibility, digestibility simulation, and development of efficient systems (Okolie et al., 2019).

However, the extraction method is a key step to obtain final products with acceptable physicochemical properties, and the so-called green technologies have diversified due to the need for using more ecofriendly processes.

Conventional extraction is where solvents or a mixture of solvents are used for homogenizing the raw material with the application of constant stirring with or without heating so that the extraction is carried out by diffusion and mass transfer. However, the existing simple techniques that do not require sophisticated equipment is unfortunately not efficient (Barba et al., 2017).

Among the conventional extraction methods are infusion, decoction, maceration, percolation, reflux, and Soxhlet (Panja, 2018).

Unconventional techniques require energy during the process for enhancing the extraction yield and time; additionally, these techniques are more selective (Barba et al., 2017).

Among the most common techniques are the following:

- *Enzyme-assisted extraction:* It is effective for the extraction of polyphenols bound to proteins and carbohydrates; the most commonly used enzymes are cellulose, pectinase, and protease.

- *UAE:* It is based on the phenomenon of cavitation with the presence of bubbles, which grow and disrupt, allowing the collapse of raw materials and the entry of solvents. Generally used for extraction of oils and polyphenols, it is mainly based on the ultrasound energy to facilitate the release of organic and inorganic substances in the matrix of the plant by increasing the mass transfer and accelerating the access of the solvent to the intracellular content (Zhu et al., 2016).
- *Microwave-assisted extraction:* This involves irradiating the sample immersed in the solvent, where ultrasonic waves direct the heating through the raw material and the solvent. Mainly used at the laboratory level, it reduces energy consumption and the amount of the solvent, thereby decreasing the amount of waste (Lefsih et al., 2017). Its main advantage is the extraction yield, by improving the mass transfer, cell disruption, and solvent penetration (Vinatoru et al., 2017).
- *Supercritical fluid extraction:* It is defined by temperature and pressure. It is based on the use of the dissolving capacity of the fluids at a temperature and pressure above the critical levels; therefore, their properties are intermediate between the liquid and gas, increasing their ability to penetrate the solid material. The fluid most commonly used is CO_2 because it is cheap, available in high quantity, flammable, and chemically inert (Panja, 2018).
- *Electrical pulses:* It is a nonthermal process; the treatment is applied to a material placed between high-voltage electrodes, in short periods of time (less than 1 s) (Zhu et al., 2016).

It is very efficient in extracting polyphenols from natural sources (Panja, 2018).

Some of these techniques have already been used for the extraction of bioactive compounds from the nopal plant; some of the unconventional techniques used in the residues generated from the fruit of different species of *Opuntia* have been summarized, such as the supercritical extraction with CO_2 to obtain oil, microwave extraction for obtaining pectin and mucilage, and electrical pulses for extracting natural dyes (Barba et al., 2017).

Each method has its advantages and disadvantages; the key is to choose the appropriate method for extracting the required compound is to consider the physical characteristics of the raw material used. As mentioned earlier, bioactive compounds that can be obtained will depend on the variety or the species of *Opuntia*.

KEYWORDS

- **physicochemical properties**
- **extraction methodologies**
- **agroindustrial wastes**
- **bioactive compounds**
- **food industries**

REFERENCES

Akinhanmi, T.F., Atasie, V.N., Akintokun, P.O. (2008). Chemical composition and physicochemical properties of cashew nut (*Anacardium occidentale*) oil and cashew nut shell liquid. Journal of Agricultural Food and Environment Science, 2, 1–10.

Álvarez-Casas, M., García-Jares, C., Llompart, M., Lores, M. (2014). Effect of experimental parameters in the pressurized solvent extraction of polyphenolic compounds from white grape marc. Food Chemistry, 157, 524–532.

Ameer, K., Bae, S-W., Jo, Y., Lee, H-G., Ameer, A., Kwon, J-H. (2007). Optimization of microwave-assisted extraction of total extract, stevioside and rebaudioside-A from *Stevia rebaudiana* (Bertoni) leaves, using response surface methodology (RSM) and artificial neural network (ANN) modelling. Food Chemistry, 229, 198–207.

Ammar, I., Ennouri, M., Attia, H. (2015). Phenolic content and antioxidant activity of cactus (*Opuntia ficus-indica* L.) flowers are modified according to the extraction method. Industrial Crops and Products, 64, 97–104.

Andrade, T., Araújo, B.Q., Citó, A.M., da Silva, J., Saffi, J., Richter, M.F., Ferraz, A., de Barros. A. (2011). Antioxidant properties and chemical composition of technical cashew nut shell liquid (tCNSL). Food Chemistry, 126(3), 1044–1048.

Andreu, L., Nuncio-Jáuregui, N., Carbonell-Barrachina, Á., Legua, P., Hernández, F. (2018). Antioxidant properties and chemical characterization of Spanish *Opuntia ficus-indica* Mill. cladodes and fruits. Journal of the Science of Food and Agriculture, 98(4), 1566–1573.

Barba, F., Putnik, P., Bursać, D., Poojary, M., Roohinejad, S., Lorenzo, J., Koubaa, M. (2017). Impact of conventional and non-conventional processing on prickly pear (*Opuntia* spp.) and their derived products: From preservation of beverages to valorization of by-products. Trends in Food Science and Technology, 67, 260–270.

Barba, F.J., Zhenzhou Z., Mohamed, K., Anderson, S. (2016). Green alternative methods for the extraction of antioxidant bioactive compounds from winery wastes and by-products: A review. Trends in Food Science and Technology, 49, 96–109.

Cheikh, M., Abdelmoumen, S., Thomas, S., Attia, H., Ghorbel, D. (2018). Use of green chemistry methods in the extraction of dietary fibers from cactus rackets (*Opuntia ficus-indica*): Structural and microstructural studies. International Journal of Biological Molecules, 116, 901–910.

Chemat, F., Rombaut, N., Sicaire, A., Mullemiestre, A., Fabiano-Tixier, A-S., Abert-Vian, M. (2017). Ultrasound assisted extraction of food and natural products. mechanisms, techniques, combinations, protocols and applications. A review. Ultrasonics Sonochemistry, 34, 540–560.

Cholake, S.T., Rajarao, R., Henderson, P., Rajagopal, R.R., Sahajwalla, V. (2017). Composite panels obtained from automotive waste plastics and agricultural macadamia shell waste. Journal of Cleaner Production, 15, 163–171.

Chupin, L., Maunu, S.L., Reynaud, S., Pizzi, A., Charrier, B., Charrier-El Bouhtoury, F. (2015). Microwave assisted extraction of maritime pine (*Pinus Pinaster*) bark: Impact of particle size and characterization. Industrial Crops and Products, 65, 142–149.

Cortés, S., Rodríguez, R., Salgado, J. M., Domínguez, J. M. (2011). Comparative study between Italian and Spanish grape marc spirits in terms of major volatile compounds. Food Control, 22, 673–680.

Diaz A. B., Blandino, A., Caro, I. (2018). Value added products from fermentation of sugars derived from agro-food residues. Trends in Food Science & Technology, 71, 52–64.

Dorantes-Alvarez, L., Miñon-Hérnandez D., Ordaz-Trinidad, N., Guzman-Geronimo, R. (2017). Microwave-assisted extraction of phytochemicals and other bioactive compounds. Reference Module in Food Science, 1–10. DOI (10.1016/B978-0-08-100596-5.21437-6)

Ekezie, F., Da, W., Jun H. (2017). Acceleration of microwave-assisted extraction processes of food components by integrating technologies and applying emerging solvents: A review of latest developments. Trends in Food Science and Technology, 67, 160–172.

Eleojo, C., Amoo, S., Kudanga, T. (2018). *Opuntia* (Cactaceae) plant compounds, biological activities and prospects—A comprehensive review. Food Research International, 112, 328–344.

Eleojo-Aruwa, C., Ammo, S., Kudanga, T. (2019). Phenolic compound profile and biological activities of Southern African *Opuntia ficus-indica* fruit pulp and peels. LWT—Food Science and Technology, 111, 337–344.

Fan, F., Yang, Z., Li, H., Shi, Z., Kan, H. (2018). Preparation and properties of hydrochars from macadamia nut shell via hydrothermal carbonization. Royal Society Open Science, 5(10), 181126.

FAO. (2018). Ecología del cultivo, manejo y usos del nopal. Roma, Italia: Organización de las Naciones Unidas para la Alimentación y la Agricultura y el Centro Internacional de Investigaciones Agrícolas en Zonas Áridas. Retrieved from http://www.fao.org/3/i7628es/I7628ES.pdf.

Felkai-Haddache, L., Dahmoune, F., Remini, H., Lefsih, K., Mouni, L., Madani, K. (2016). Microwave optimization of mucilage extraction from *Opuntia ficus indica* cladodes. International Journal of Biological Macromolecules, 84, 24–30.

García-Cayuela, T., Gómez-Maqueo, A., Guajardo-Flores, D., Welti-Chanes, J., Cano, M. (2019). Characterization and quantification of individual betalain and phenolic compounds in Mexican and Spanish prickly pear (*Opuntia ficus-indica* L. Mill) tissues: A comparative study. Journal of Food Composition and Analysis, 76, 1–13.

Garrido, T., Gizdavic-Nikolaidis, M., Leceta, I., Urdanpilleta, M., Guerrero, P., de la Caba, K., Kilmartin, P. (2019). Optimizing the extraction process of natural antioxidants from chardonnay grape marc using microwave-assisted extraction. Waste Management, 88, 110–117.

Guandalini, V, Rodrigues, N. Ferreira M.L. (2019). Sequential extraction of phenolics and pectin from mango peel assisted by ultrasound. Food Research International, 119, 455–461.

Guerra-Olgin, R. (2012). Evaluación de Extractos Polifenólicos de Residuos de Nogal Pecanero (*Carya illinoensis*) Obtenidos Mediante Técnicas de Extracción Alternativas Para Su Efecto Contra Bacterias Patógenas a Humanos. Thesis. Universidad Autónoma Agraria Antonio Narro, Buenavista, Saltillo, Coahuila, México.

Hamad, F.B., Mubofu, E.B. (2015). Potential biological applications of bio-based anacardic acids and their derivatives. International Journal of Molecular Sciences, 16(4), 8569–8590.

Hilbig, J., Alves, V.R., Müller, C.M.O., Micke, G.A., Vitali, L., Pedrosa, R.C., Block, J.M. (2018). Ultrasonic-assisted extraction combined with sample preparation and analysis using LC-ESI-MS/MS allowed the identification of 24 new phenolic compounds in pecan nut shell [*Carya illinoinensis* (Wangenh) C. Koch] extracts. Food Research International, 106, 549–557.

Kaderides, K., Papaoikonomou, L., Serafim, M., Goula, A. Microwave-assisted extraction of phenolics from pomegranate peels: Optimization, kinetics, and comparison with ultrasounds extraction. Chemical Engineering and Processing—Process Intensification, 137, 1–11.

Khadhraoui, B., Turk, M., Fabiano-Tixier, A.S., Petitcolas, E., Robinet, P., Imbert, R., Maataoui, M.E., Chemat, F. (2018). Histo-cytochemistry and scanning electron microscopy for studying spatial and temporal extraction of metabolites induced by ultrasound. towards chain detexturation mechanism. Ultrasonics Sonochemistry, 42, 482–492.

Lefsih, K., Giacomazza, D., Dahmoune, F., Mangione, M., Bulone, D., San Biagio, P., Madani, K. (2017). Pectin from *Opuntia ficus indica*: Optimization of microwave-assisted extraction and preliminary characterization. Food Chemistry, 221, 91–99.

Lomonaco, D., Santiago, G.M.P., Ferreira, Y.S., Arriaga, A.M.C., Mazzetto, S.E., Mele, G. (2009). Study of technical CNSL and its main components as new green larvicides. Green Chemistry, 11: 31–33.

Lubi, M.C., Thachil, E.T. (200). Cashew nut shell liquid (CNSL)—A versatile monomer for polymer synthesis. Designed Monomers and Polymers, 3, 123–153.

Marin-Bustamante, M., Chanona-Pérez, J., Guemes-Vera, N., Arzate-Vázquez, I., Perea Flores, M., Mendoza-Pérez, J., Cásarez-Santiago, R. (2018). Evaluation of physical, chemical, microstructural and micromechanical properties of nopal spines (*Opuntia ficus-indica*). Industrial Crops and Products, 123, 707–718.

Marin-Bustamante, M., Chanona-Pérez, J., Güemes-Vera, N., Cásarez-Santiago, R., Perea-Flores, M., Arzate-Vázquez, I., Calderón-Domínguez, G. (2017). Production and characterization of cellulose nanoparticles from nopal waste by means of high impact milling. Procedia Engineering, 200, 428–433.

Melgar, B., Dias, M., Ciric, A., Sokovic, M., García-Castello, E., Rodríguez-López, A., Fereira, I. (2017). By-product recovery of *Opuntia* spp. peels: Betalainic and phenolic profiles and bioactive properties. Industrial Crops and Products, 107, 353–359.

Mlowe, S.S., Pullabhotla, R.R., Mubofu, E. B., Ngassapa, F.N., Revaprasadu, N.N. (2014). Low temperature synthesis of anacardic-acid-capped cadmium chalcogenide nanoparticles. International Nano Letters, 4, 106–111.

Okolie, C., Akanbi, T., Mason, B., Udenigwe, C., Aryee, A. (2019). Influence of conventional and recent extraction technologies on physicochemical properties of bioactive macromolecules from natural sources: A review. Food Research International, 116, 827–839.

Pandit, S., Pasupuleti, V., Kulkarni S.G. (2015). Pectic principles of mango peel from mango processing waste as influenced by microwave energy. LWT—Food Science and Technology, 64(2), 1010–1014.
Panja, P. (2018). Green extraction methods of food polyphenols from vegetable materials. Current Opinion in Food Science, 23, 173–182.
Parkash-Maran, J., Swathi, K., Jeevitha, P., Jayalakshmi, J., Ashvini, G. (2015). Microwave-assisted extraction of pectic polysaccharide from waste mango peel. Carbohydrate Polymers, 123, 67–71.
Pérez-Méndez, L., Tajera-Flores, F., Darias, J., Rodríguez, E., Díaz, C. (2015). Physicochemical characterization of cactus pads from *Opuntia dillenii* and *Opuntia ficus indica*. Food Chemistry, 188, 393–398.
Pinheiro do Prado, A., Silvestre da Silva, H., Mello da Silveira, S., Manique Barreto, P.L., Werneck Vieira, C.R., Maraschin, M., Salvador Ferreira, S.R., Mara Block, J. (2014). Effect of the extraction process on the phenolic compounds profile and the antioxidant and antimicrobial activity of extracts of pecan nut. Industrial Crops and Products, 52, 552–561.
Pinheiro do Prado, A.C., Manion, B., Seetharaman, K., Deschamps, F.C., Barrera, D., Mara-Block, J. (2013). Relationship between antioxidant properties and chemical composition of the oil and the shell of pecan nuts [*Carya illinoinensis* (Wangenh) C. Koch]. Industrial Crops and Products, 45, 64–73.
Poinern, G., Senanayake, N., Shah, X.N., Thi-Le, G.M., Parkinson, D. (2011). Adsorption of the aurocyanide complex on granular activated carbons derived from macadamia nut shells—A preliminary study. Minerals Engineering, 24(15), 1694–1702.
Rajarao, R., Mansuri, I., Dhunna, R., Khanna, S.V. (2014). Study of structural evolution of chars during rapid pyrolysis of waste CDs at different temperatures. Fuel, 134, 17–25.
Richard, T., Saeid, B. (2017). A techno-economic comparison of subcritical water, supercritical CO_2 and organic solvent extraction of bioactives from grape marc. Journal of Cleaner Production, 158, 349–358.
Rodrigues, L.A., Thim, G.P., Ferreira, R.R, Alvarez-Mendez, M.O., Coutinho, A. (2013). Activated carbon derived from macadamia nut shells: An effective adsorbent for phenol removal. Journal of Porous Materials, 20, 619–627.
SAGARPA. (2017). Uva mexicana. México: Subsecretaría de Agricultura. Retrieved from https://www.gob.mx/cms/uploads/attachment/file/257085/Potencial-Uva.pdf.
Salehi, E., Emam-Djomeh, Z., Askari, G., Fathi, M. (2019). *Opuntia ficus indica* fruit gum: Extraction, characterization, antioxidant activity and functional properties. Carbohydrate Polymers, 206, 565–572.
Sant´Anna, V., Brandelli, A., Damasceno, L., Cristina, I. (2012). Kinetic modeling of total polyphenol extraction from grape marc and characterization of the extracts. Separation and Purification Technology, 100, 82–87.
Schmidt, C.G., Gonçalves, L.M., Prietto, L., Hackbart, H.S., Furlong, E.B. (2014) Antioxidant activity and enzyme inhibition of phenolic acids from fermented rice bran with fungus *Rizhopus oryzae*. Food Chemistry, 146, 371–377.
Sólyom, K., Kraus, S., Mato, R., Gaukel, V., Schuchmann, H., Cocero, M. (2013). Dielectric properties of grape marc: Effect of temperature, moisture content and sample preparation method. Journal of Food Engineering, 119, 33–39.
Stasiuk, M., Kozubek, A. (2010). Biological activity of phenolic lipids. Cellular and Molecular Life Sciences, 67,, 841–860.

Tamires, D., Gadioli, A., Baú, C., Fernández, G., Martínez, J. (2019). Pressurized liquid extraction of bioactive compounds from grape marc. Journal of Food Engineering, 240, 105–113.

Tiwari, B.K. (2015). Ultrasound: A clean, green extraction technology. TrAC—Trends in Analytical Chemistry, 71, 100–109.

Torres-Ponce, R., Morales-Corral, D., Ballinas-Casarrubias, L., Nevárez-Morillón, G. (2015). Nopal: Semi-desert plant with applications in pharmaceuticals, food and animal nutrition. Revistsa Mexicana de Ciencias Agrícolas, 6(5), 1129–1142.

Vinatoru, M., Mason, T., Calinescu, I. (2017). Ultrasonically assisted extraction (UAE) and microwave assisted extraction (MAE) of functional compounds from plant materials. TrAC—Trends in Analytical Chemistry, 97, 159–178.

Vinatoru, M., Mason, T., Calinescu, I. (2017). Ultrasonically assisted extraction (UAE) and microwave assisted extraction (MAE) of functional compounds from plant materials. TrAC—Trends in Analytical Chemistry, 97, 159–178.

Wen, C., Zhang, J., Zhang, Sedem, C., Zandile, M., Duan, Y., Ma, H., Luo, X. (2018). Advances in ultrasound assisted extraction of bioactive compounds from cash crops—A review. Ultrasonics Sonochemistry, 48, 538–549.

Xu, M., H., Yang, X., Fu. M.R. (2016). Combined ultrasonic and microwave method for juglone extraction from walnut green husk (*Juglans nigra*). Waste Biomass Valorization, 7(5), 1159–1166.

Yang, T., Zhihang, Z., Da-Wen, S. (2014). Kinetic modeling of ultrasound-assisted extraction of phenolic compounds from grape marc: Influence of acoustic energy density and temperature. Ultrasonics Sonochemistry, 21, 1461–1469.

Zhu, Z., He, J., Liu, G., Barba, F., Koubaa, M., Ding, L., Vorobiev, E. (2016). Recent insights for the green recovery of inulin from plant food materials using non-conventional extraction technologies: A review. Innovative Food Science and Emerging Technologies, 33, 1–9.

CHAPTER 3

PHYSICOCHEMICAL PROPERTIES OF PRODUCTS MADE FROM MIXTURES OF CORN AND LEGUMES

DANIELA SÁNCHEZ-ALDANA VILLARRUEL*,
MARTHA YARELI LEAL RAMOS, TOMÁS GALICIA GARCÍA,
RUBEN MARQUEZ MELÉNDEZ, and RICARDO TALAMÁS ABBUD

Facultad de Universidad Autónoma de Chihuahua, Circuito Universitario S/N, Campus Uach II, Chihuahua 31125, Chih., México

Corresponding author. E-mail: dsancheza@uach.mx

ABSTRACT

Physicochemical properties play an important role in maintaining the quality of the products made from mixtures of corn and legumes that are fundamental in the research and development of new products, the design of equipment, the improvement of processes, and quality control of raw materials, intermediates, and finished products. The physicochemical properties are intimately related to the functional properties of the constituents of the food system, as well as the operational variables that are applied in the different stages of the process. Likewise, the nixtamalization of maize is a thermal-alkaline treatment that is responsible for important changes in the physicochemical, thermal, nutritional, and sensory characteristics of flours, doughs, and finished products. In this chapter, the physicochemical properties of different products made from mixtures of corn and legumes products such as flours, doughs, bakery, tortillas, snacks, and breakfast cereals will be reviewed.

3.1 INTRODUCTION

Corn (*Zea mays*), also referred to as maize, is among the most extensively cultivated and consumed cereal crops in the world and used for the production of an array of human food, animal feeds, biofuels, and other industrial items (Tazrat et al., 2019). It is the main cereal grain as measured by production but ranks third as a staple food, after wheat and rice (Tazrat et al., 2019; Gwirtz and Garcia-Casal, 2013; Kaur et al., 2015). Currently, the United States, Brazil, Mexico, Argentina, India, France, Indonesia, South Africa, and Italy produce 79% of the world's maize production (Gwirtz and Garcia-Casal et al., 2013; Abebe and Chandravanshi, 2017). Corn flour (CF) is one of the most used raw materials as the main ingredient for the production of gluten-free (GF) products due to its abundance, low cost, and absence of gluten (Gao et al., 2017; Gimenez et al., 2013; Palavecino et al., 2017). However, it is low in protein and dietary fiber contents.

On the other hand, legumes are the second major source for humans next to cereals and play an important role in the human diet in developing countries. Legumes are rich in proteins (20–37%, w/w), easily available and rich in dietary fiber (3–31% w/w) and resistant starch (RS, 11–20% w/w) (Guiberti et al., 2015; Laleg et al., 2017). Legume flours, concentrates, or isolates are a good supplement for cereals because they not only increase protein content from 10–12% to greater than 15% but also improve the biological value of the protein (chemical score from 45 to >70) (Marconi and Messia, 2012; Elias, 1996).

Nowadays, the industry of products based on wheat such bread, cookies, pasta, tortillas, snacks, breakfast cereals, supplements, among others produces products based not only including wheat in formulation but also on other sources of vegetable origin, either forming a majority part of the formulation or in a complementary way with wheat flour (WF). The flours used to make these alternative formulations are known as compound flours, classified as flours without wheat (from other cereals, legumes, and tubers) and diluted WFs where a part of the WF is substituted (Elias, 1996). Gómez et al (2013) indicated that one of the causes of the partial or total replacement of WF is due to the high demand of this cereal worldwide where the shortage leads to the need to search for alternative formulations, while García (2006) indicated that part of the world population (3%) has a gastrointestinal pathology where there is an intolerance toward gluten proteins, known as celiac disease, so that the food industry in some products has excluded ingredients like wheat and some cereals that contain gluten (oats, rye, and barley); hence, the use of legumes and cereals such as corn is the subject of study in the development of new products (Umaña et al., 2013).

In the present chapter, we will review the physicochemical properties of finished products made from mixtures of corn and legumes.

3.2 PHYSICOCHEMICAL PROPERTIES OF BAKERY PRODUCTS

One of the most important products in human nutrition is represented by bread products because it has functional ingredients such as fiber or the incorporation of micronutrients into flours, which together with its availability and low cost makes it a staple food in the daily diet (Umaña et al., 2013).

In the bakery, the physiochemical properties that are important to evaluate are color, volume, the structure of the bread crumb, and texture profile such as hardness, elasticity, adhesiveness, cohesiveness, fragility, chewiness and gumminess, and penetration in biscuits (Sanchez-Hoyos and Peña-Palacios, 2015).

The color of a bakery product, in the first instance, provides information about its appearance, which is undoubtedly a decisive feature in consumer preference. Cauvin and Young (2006) report that the color evaluation in bakery products is done externally and internally, so the external color depends directly on the proportion of each ingredient as well as the baking, while internally one of the aspects that influences more is the alveolar structure of the crumb.

Bread volume is one of the most evaluated characteristics in bread since it indicates its external quality together with other evaluations such as its appearance, color, dimensions, as well as its crust (Cauvin and Young, 2002). The bread volume is directly influenced by the ingredients used in its formulation and manufacture, being the seed displacement technique the most used for its determination (Moreira et al., 2007), where once the seed to be used is standardized, the piece is introduced into a container and filled with seeds, displacing a volume corresponding to the sample. Another of the complementary analyses is the evaluation of the alveolar structure of the crumb, which is directly related to the volume and is an indicator of the internal quality of the bread; its estimation is carried out subjectively with a standard, but due to the different characteristics present in commercial products (bread type), there is not a single standard for all products.

The firmness and elasticity are two mechanical properties of interest in the evaluation of bread crumbs; many of the existing analyses are based on sensory parameters such as chewiness and palatability (Roudot, 2004). One of these analyses corresponds to the texture profile analysis (TPA) performed by two compression cycles in texture equipment to imitate the action of an individual's jaw on the food. By means of the graphical representation of the test, a force–time curve is generated by obtaining the seven textural parameters. Rosenthal (2001) describes and lists the following properties:

1. *Hardness:* Force to compress a food between the molars.
2. *Elasticity:* Extension to which a compressed food returns to its original size once the applied force is removed.
3. *Adhesiveness:* Work necessary to remove food from the surface.
4. Cohesiveness: Strength of internal links on food.
5. *Fragility:* Force with which the material fractures.
6. *Chewiness:* Energy required to chew a food prior to being swallowed. This parameter is determined by the product of hardness, cohesiveness, and elasticity.
7. *Rubbery:* Energy required to disintegrate a semi-solid food ready to be swallowed. Its determination is made between the product of hardness and cohesiveness.

In the case of cookies or products where a low volume is obtained, a hardness analysis is carried out in which it is intended to simulate the penetration of the teeth on a solid food so that its analysis is carried out in texture equipment coupled to a probe (shape and known size), which is introduced and evaluates the force necessary to perform the penetration.

3.2.1 PHYSICOCHEMICAL CHARACTERIZATION OF BAKERY PRODUCTS BASED ON CORN AND LEGUMES

Cornbread and cookie products have the characteristic of lacking the protein fraction responsible for imparting mass functional characteristics such as elasticity and elongation; however, despite this limitation by incorporating hydrocolloids, pseudocereals and legumes have substantially improved the functionality of the final product. Added to this, the presence of a GF product has made these to be accepted by people with problems of assimilation of gluten (celiac disease), which represents a potential market for high growth and added value. Below are several studies conducted on corn and legumes for the preparation of bread and cookie products, as well as their characterization of their main physicochemical properties.

Buresova et al. (2014) report the rheological study in GF doughs based on cornmeal, amaranth, chickpea, millet, quinoa, and rice and their effect on the quality of yeast bread. The results obtained from the uniaxial deformation tests in doughts exhibit a greater resistance to extension, extensibility, and greater stress at the time of rupture that are directly related to obtain a better quality bread.

Sanchez-Hoyos and Peña-Palacios (2015) developed two bread and one cookie products using CF, WF, chickpea flour (CHF), and pineapple fiber mixtures. The formulations are raised as follows: (a) bread (P): 25% HM/57% HT/15% HG/3% FP and (b) cookie (G): 45% HM/20% HT/35% HG/10% FP. Based on their physicochemical characterization, it was obtained that the loaves obtained from the compound formulation (P) presented a volume of 700 ± 10 mL unlike the control (100% HT) 1350 ± 13 mL, presenting an opposite effect to the increase the hardness values in P (17.12 ± 0.46 N) compared to the target (8.53 ± 0.29 N). So, it was concluded that a low substitution in both products (bread 33% and 80% in cookies) has better characteristics and resemblance to control. The development of this product represents the potential in obtaining GF and fiber-enriched products.

Gómez et al. (2008) report the incorporation of CHF in the formulation of two types of cake, where it is concluded that the physicochemical parameters evaluated (volume, symmetry, chroma, L^*, hardness) in crust and crumb decreased when the concentration of CHF in the formulation increased, with an increase in adhesiveness. So, the incorporation of legumes in bakery products decreased their physicochemical properties; however, the nutritional contribution is important unlike a product from wheat.

Pérez-Hernandez et al. (2018) studied the chemical composition of white bread from WF, legume (bean flour BF) and oilseed mixture (sunflower-GF, sesame-AF) in a proportion (70/15/15, WF/BF/GF-AF, respectively). In this study, it is concluded that by incorporating legume and oilseed the protein content increased (9.91% WF/BF/GF and 9.86%

WF/BF/AF) unlike the control (8.90% WF), as well as an increase up to 76% in the content of Ca and 33% of P with respect to that of the control.

3.3 PHYSICOCHEMICAL PROPERTIES OF PASTA

Pasta is a key element in the basic diet of most cultures and is, consequently, one of the most popular and widely consumed foodstuffs all over the world (Das et al., 2017; Madeira Moreira da Silva et al., 2016).

The ease of pasta manufacturing and the simplicity of the raw material formulation make pasta a relatively cheap food product to manufacture, consequently with a relatively low cost for the consumer (Carini et al., 2014). There are various types of pasta, with dried pasta being appreciated by the consumers because of its convenience of use and long shelf-life (Aravind et al., 2011; Dib et al., 2018; Salch et al., 2016). Pasta is a product of easy handling, storage, and preparation (Madeira Moreira da Silva et al., 2016). Its wide diversity of forms and sensory appeal (odor, taste, and texture) has made it to be recognized as a very versatile dish from the gastronomic point of view (Cole, 1991; Fuad and Prabhasankar, 2010; Shobba et al., 2015; Sicignano et al., 2015).

The visual appearance of pasta (uncooked and cooked) and its behavior during cooking are the most important attributes that define pasta quality (Sissons et al., 2012; Turnbull, 2001). Pasta with a good appearance has a combination of physicochemical properties that include a uniform color (which implies the absence of streaks and white spots), lack of cracks/fissures, a homogenous surface texture, and symmetrical dimensions (Li et al., 2014).

The evaluation of quality in cooked pasta is related to the physicochemical properties such as cooked weight, cooking loss (CL), total organic matter (TOM), and textural properties (firmness, stickiness, elasticity, and bulkiness). Optimum cooking time (OCT), corresponding to the disappearance of the white color in the central core of a piece of pasta, should be included among cooked pasta properties. Cooking physicochemical properties can be determined by sensorial, instrumental, and chemical tests. Sensory analysis using highly trained panelists is considered the ultimate tool for the measurement of the cooking quality of pasta products.

Cooked weight is measured as the water-absorbing capacity (WAC) of pasta products during cooking, that is, the increase in weight or volume of pasta after cooking. Normally, the cooked weight is about three times the dry weight of the spaghetti.

On the other hand, CL is determined as the percentage of solids lost in the cooking water (evaluated as the solid residue in cooking water). CL, that is, leaching of starch into the cooking water, should be minimal for acceptable pasta quality. For a good quality spaghetti, the residue should not exceed 7%–8% of the dry weight of the pasta.

TOM is the material released from rinsed cooked pasta. This method assumes that pasta stickiness results from substances escaping from the protein network and adhering to the surface of cooked pasta. The main constituent of TOM is starch, mainly amylopectin, which is responsible for the stickiness of pasta. Generally speaking, TOM values >2.1 g/100 g corresponds to low quality, values between 2.1 and 1.4 g/100 g predict good quality, and values <1.4 g/100 g indicate very good quality.

Firmness is the initial resistance to penetration offered by the cooked pasta as it is chewed or flattened between the fingers or sheared between the teeth. The firmness of a cooked pasta sample is usually determined as the maximum force recorded in a compression/cut test to fracture or to reach a certain deformation of the sample.

Stickiness/adhesiveness is the state of surface disintegration of cooked pasta (Sissons et al., 2012). It is generally measured by touching/deforming the pasta sample in a controlled manner and then assessing the force needed to separate the probe from the sample. Also, the stickiness can be estimated by visual inspection, with or without the aid of standard reference pasta (Sissons et al., 2012). In other words, stickiness is the overall mouthfeel of the pasta together with any residual starch left in the mouth after swallowing (Turnbull, 2001).

Elasticity is how the pasta breaks down in the mouth on further chewing (Turnbull, 2001) and is usually determined as the tensile force to break the sample (Carini et al., 2014).

Finally, bulkiness is the degree of adhesion of pasta pieces after cooking and is evaluated visually and manually. Bulkiness is strictly correlated with stickiness (Turnbull, 2001).

The raw material characteristics and processing conditions during pasta production play a key role in determining the physicochemical properties and palatability of final pasta products and therefore their quality (De la Peña and Manthey, 2014).

Durum wheat (*Triticumturgidum L.* var. *Durum*) semolina is ideally suited for the production of high-quality pasta with good cooking behavior, stability to overcooking, and unmatchable smell and taste (Sicignano et al., 2015).

Traditional wheat pasta is known as one of the simplest food products in terms of ingredients and processing (Marti and Pagani, 2013; Mirhosseini et al., 2015). Pasta products are composed mainly of two basic ingredients: coarse semolina of durum wheat and water (Padalino et al., 2014). Occasionally, other raw materials, such as (1) protein supplements, (2) emulsifiers and edible gums, (3) antioxidants and preservatives, and (4) various plant-based or animal-based supplements, are added (Li et al., 2014). All of the other ingredients added are just optional and can be of greater or lesser importance (Shobba et al., 2015). Semolina and water are mixed into a crumble dough before being formed by extrusion or sometimes lamination (Li et al., 2014) and cut into appropriate shapes (Joytsna et al., 2014). The product is then either sold as a fresh (wet) commodity or can be dried for future use (Padalino et al., 2014).

Although wheat gluten proteins (gliadins and glutenins), which are water-insoluble storage proteins, form only a small part of the endosperm (~7%–20%), they constitute 85% of endosperm proteins (Sicignano et al., 2015). Starch presents up to 80% of semolina dry matter (Tazrart et al., 2019).

In wheat pasta, protein quantity and quality (e.g., gluten) are the most important factors affecting pasta cooking properties (Padalino et al., 2014). High protein content and gluten strength (in terms of its viscoelasticity) are required to process semolina into a suitable final pasta product with an optimal cooking performance (Marti and Pagani et al., 2013). It is accepted that weak and inelastic gluten produces pasta with low cooking quality. The structure and amylose content of starch granules were found to influence pasta characteristics (Sicignano et al., 2015).

In uncooked dried pasta, the gluten network is more or less uniformly and regularly arranged around starch granules according to the quality of the semolina used and starch is still in the form of whole native granules, as in semolina (Marti and Pagani et al., 2013).

During cooking, starch and protein exhibit completely different behaviors (Gao et al., 2017). The starch granules swell and partly solubilize during cooking, while proteins become completely insoluble and coagulate, creating a strengthened network, which traps the starch material (De la Peña and Manthey, 2014). Starch gelatinization and protein coagulation are both competitive phenomena, occurring at the same temperature and are influenced by water availability (Marti and Pagani, 2013). "Good-quality" cooked pasta should have a firm texture to the bite (*al dente*), be nonsticky, and have a minimal loss of solids in the cooking water (Carini et al., 2014). The faster the formation of a continuous protein network, the more limited the starch swelling, reducing the loss of solids in the cooking water and thus ensuring firm consistency and the absence of surface stickiness in pasta (Marti and Pagani, 2013). On the contrary, if the protein network lacks elasticity or its formation is delayed, starch granules will easily swell and part of the starchy material will pass into the cooking water, resulting in a product characterized by stickiness and poor consistency (Marti and Pagani, 2013).

There are many reasons for using nontraditional raw materials such as corn and legumes in pasta making. Among the main ones is the production of GF pasta with improved protein content and quality as well as increased dietary fiber (Padalino et al., 2014; Camelo-Méndez et al., 2018; Flores-Silva et al., 2014). Consumers with celiac disease mainly demand GF pasta (O´Shea et al., 2014).

Celiac disease is a lifelong intolerance to the prolamins (gliadins) present in wheat (i.e., all *Triticum* species, such as durum wheat, spelt, and kamut), the prolamins of rye (secalins), barley (hordeins), and possibly oats (avidins), or their crossbred derivatives (Larrosa et al., 2015). Celiac disease is characterized by immune-mediated damage to the gut mucosa, which may affect the absorption of important nutrients such as iron, folic acid, calcium, and fat-soluble vitamins (Larrosa et al., 2015). The only effective treatment for celiac disease is strict adherence to a GF diet during the patient's lifetime (Padalino et al., 2014).

3.3.1 PHYSICOCHEMICAL PROPERTIES OF PASTA MADE FROM MIXTURES OF CORN AND LEGUMES

Pasta obtained from nonconventional flours should emulate the characteristics of traditional pasta products, such as color, cooking properties, texture (including elasticity, firmness, and reduced adhesiveness), and taste (Flores-Silva et al., 2014).

However, the preparation of pasta from unconventional ingredients that are GF is difficult because of the lack of gluten, which contributes to the development of a strong protein network that prevents the dissolution of starch from pasta during cooking. In addition, nontraditional ingredients could modify the rheological properties of the dough or the sensory acceptability and cooking quality of the pasta. Therefore, balanced formulations and adequate technological processes must be adopted to produce acceptable GF pasta (Morreale et al., 2019).

While gluten proteins play a key role in conventional semolina pasta properties, starch is the determining component in GF pasta (Marti and Pagani et al., 2013). In GF pasta, starch is responsible for the products' structure, which is related to the tendency of its macromolecules to re-associate and interact after gelatinization, resulting in newly organized structures that retard further starch swelling and solubilization during cooking. (Merayo et al., 2011). The ideal starch for GF pasta products should have a marked tendency to retrograde: this property, generally observed in high amylose cereals and legumes, assures good cooking behavior in terms of texture and low CL, even after prolonged cooking (Marti and Pagani, 2013).

The common ingredients in GF pasta made from blends of corn and legumes are CF and/or isolated starch, with the addition of protein from legumes flours, hydrocolloids or gums, and emulsifiers, which may partially act as substitutes for gluten (Merayo et al., 2011). Today, GF starchy flours are used more than isolated starches, thus skipping the expensive stage of starch extraction from the grains. Furthermore, from a technological point of view, the use of flours allows exploiting the presence of interactions between starch and other components, such as proteins and lipids (Marti and Pagani et al., 2013).

Basically, in GF pasta, the role of gluten could be replaced by choosing suitable formulations using heat-treated flours as the key ingredients, or by adopting nonconventional pasta-making processes to induce new rearrangements of starch macromolecules (Marti and Pagani et al., 2013).

Pasta prepared only from nongluten flour is generally considered to be inferior in textural quality compared to semolina pasta: it does not tolerate overcooking, it is sticky, and, above all, it is characterized by relevant CLs. Adding texturing ingredients can be a simple solution for improving pasta cooking behavior by decreasing these defects (Marti and Pagani et al., 2013; Abdel-Aal, 2009).

Hydrocolloids or gums are commonly used as texturing ingredients for their ability to make a gel in little quantities, provide high consistency at room temperature, improve firmness, and give body and mouthfeel to pasta. In addition, because of their ability to bind water, gums can increase the rehydration rate of pasta. A wide range of hydrocolloids has been proposed: Arabic gum, xanthan gum, locust bean gum, carboxymethyl cellulose (CMC), etc. (Marti and Pagani et al., 2013; Abdel-Aal, 2009).

Emulsifiers (monoglycerides and diglycerides) act as lubricants in the extrusion process and provide firmer consistency and a less sticky surface, as they control starch swelling and leaching phenomena during cooking, thereby improving the texture of the final product.

Padalino and co-workers (2014) evaluated the effects of the addition of CHF and hydrocolloids on the physicochemical properties of maize-based spaghetti. Spaghetti samples were elaborated by keeping constant the proportions of heat-treated (pregelatinized) maize flour (10%), monoglycerides (1%), and water (10 L) and varying the amount of chickpea (0%, 10%, 15%, 20%, 25%, and 30%) and the rest of the maize flour.

For noncooked dry spaghetti, the increase of the CHF up to 15% caused a decrease in the score of both color and homogeneity attributes, whereas the resistance to break values of all the enriched samples was statistically similar to the control sample (0% CHF). In general, the overall quality of cooked spaghetti decreased with the increase of the CHF amount. Spaghetti samples loaded with 15% CHF showed poor elasticity, increased firmness, unpleasant color, and homogeneity, so this concentration represented the highest CHF concentration to be used. To improve the overall sensory quality of the enriched spaghetti, they added 2% of one of the following hydrocolloids: agar, guar seed flour, or pectin from citrus.

The addition of hydrocolloids to the formulation improved the quality of uncooked pasta. This was mainly due to an improvement of color and homogeneity, whereas the resistance to break was statistically similar to the maize-based spaghetti enriched with 15% of CHF without hydrocolloids. The addition of hydrocolloids also improved elasticity, homogeneity, odor, and taste of the cooked pasta. The hydrocolloids help gelatinized maize starch to form a stable network that improves pasta structure and its overall quality.

Despite the several well-known positive effects of the addition of emulsifiers and hydrocolloids, consumers often associate their presence in GF pasta to an "artificial" food (Marti and Pagani et al., 2013). To counteract the above, the use of functional ingredients such as RS and natural antioxidants can be a viable alternative to improve the healthy image of GF pasta while positively impacting its physicochemical properties.

Flores-Silva and co-workers (2014) developed GF spaghetti with a low glycemic index (GI) using mixtures of white maize, chickpea, and unripe plantain (stage 2) flours and compared its cooking quality with that of a control semolina spaghetti.

Legumes, such as chickpea, as well as unripe plantain (*Musa paradisiaca* L.), have high RS content and low starch digestion rate. RS has been associated with a slower digestion rate and therefore a slower glucose release into the bloodstream after its consumption, resulting in reduced glycemic and insulinemic postprandial responses in comparison with diets based on cereal grains or potatoes.

They prepared spaghetti samples with 100% durum wheat semolina as control and eight GF formulations with different percentages of chickpea (70%, 65%, and 60%), unripe plantain (30%, 25%, 20%, and 15%), and white maize flour (20%, 15%, 10%, 5%, and 0%). All of the GF formulations were supplemented with 0.5% of carboxymethyl cellulose.

The CL among all the GF spaghetti was in the range of 10.04–10.91% and not significantly different from each other. These values were almost at the limit of acceptability to be considered as good cooking quality.

GF spaghetti showed to have firmness, hardness, cohesiveness, and chewiness, similar or higher than semolina wheat spaghetti used as a control. Also, some physicochemical characteristics of the GF spaghetti, such as the diameter and water absorption, were similar to those of the control sample. The result of the sensory evaluation of the products concluded that the overall acceptability of the GF spaghetti was about 70% compared with the that of the control spaghetti. The use of unripe plantain, chickpea, and maize flours in the fabrication of GF spaghetti is a novel approach to provide a healthy alternative to the traditional gluten-containing pasta products to the large population of consumers suffering from coeliac disease and gluten sensitivity.

Camelo-Méndez and co-workers (2018) evaluated the physicochemical properties of GF spaghetti elaborated with white and blue maize (25%, 50%, and 75%), chickpea, and unripe plantain (stage 2) flours. All formulations contained 1.5% of CMC.

Blue maize exhibits antioxidant power and potential anticancer properties, associated with its anthocyanin content.

In general, an increase in the maize flour (white and blue) level in the pasta formulation did not affect the cooking quality variables. Although slight differences in amylose, starch, and protein contents between blue and white maize have been found, they did not have a noticeable impact on the cooking quality of the pasta samples studied. CL values, a parameter that is connected to the acceptability by consumers, were 9%–11% for GF pasta. These CL values were lower than the 12% limit suggested for GF pasta of good quality. Pasta containing blue maize flour had lower hardness, lower chewiness, and higher adhesiveness than the corresponding white maize-based samples, a pattern that is related to the lower starch content of blue maize. Blue maize-based pasta presented a dark color. The addition of blue maize flour at 75% conveyed the highest total phenolic content retention after extrusion (80%) and cooking (70%).

Up to now, GF pasta made from solely GF flour has usually been prepared in one of two ways. The first approach focuses on the use of heat-treated flours, in which starch is already mostly gelatinized. Here, the pretreated flour can be formed into pasta by the continuous extrusion press commonly used in durum wheat semolina pasta making (Marti and Pagani et al., 2013). In the second technological approach (extrusion-cooking process), native flour is treated with steam and extruded at high temperatures (more than 100 °C) for promoting starch gelatinization directly inside the extruder cooker. A careful selection of the processing conditions is the starting point for promoting new starch arrangements in GF raw materials to assure good cooking behavior and effective structure, not only for the texture but also for nutritional properties in terms of enzyme accessibility and starch digestibility (Marti and Pagani et al., 2013).

Giménez and co-workers (2013) noted that it is possible to obtain GF pasta with adequate quality characteristics without the addition of gluten substitutes such as hydrocolloids, only by exercising control in the formulation and variables of the extrusion-cooking process. Thus, they evaluated the effect of extrusion temperature (T = 80, 90, and 100 °C) and moisture content (M = 28%, 31%, and 34%) of a corn/broad bean (*Vicia faba*) flour blend in a 70:30 ratio on OCT, CL%, water absorption (WA), firmness, and stickiness of spaghetti-type pasta. The products were dried at 40 °C and 40% relative humidity for 16 h.

Giménez and co-workers (2013) found that extrusion cooking at 100 °C and 28% moisture is appropriate to obtain corn-broad bean spaghetti-type pasta with high protein and dietary fiber contents and adequate physicochemical and textural characteristics. The OCT was 13 min. Pasta extruded under those conditions had the lowest CL (9.07%), and this value increased (11.14%) under overcooking conditions (OCT + 10 min). However, this value exceeded the considered acceptable range for semolina spaghetti (7%–8%) and starch noodles (9%–10%).

This behavior may be due to the higher fiber content contributed by broad BF, which might lead to the weakening of the starch network formed.

Higher temperatures and lower moisture contents in the extrusion-cooking process allow for the creation of a more continuous and less soluble structure. The hydration properties during cooking will depend only not on the damaged and gelatinized starch but also on the resistance offered by the structures formed during extrusion.

The corn-broad bean spaghetti-type pasta obtained at 100 °C with 28% moisture presents greater restriction to water absorption (WA = 2.13 g H_2O/g pasta at OCT and WA= 2.99 g H_2O/g pasta at overcooking) during cooking. This behavior can be attributed to the formation of new structures such as retrograded amylose and amylose–lipid complexes, which favor the structural stability of corn-broad bean spaghetti, providing greater resistance to hydration with low loss of solids during overcooking.

The textural properties of pasta-like spaghetti type and their behavior during overcooking depend on the degree of gelatinization reached during the extrusion-cooking process and the formation of the amylose–lipid complex.

Sensory evaluation was carried out with a trained panel of three persons to evaluate the firmness and stickiness. The global score (GSS) was obtained by consensus in two replications. A 0–5 scale was used. The 0 value was assigned for the firm and not sticky noodles and 5 for the softer and very sticky ones. A GSS (firmness + stickiness) less than or equal to 5 was considered acceptable. The values of GSS were 2 at the OCT and 3 at overcooking.

The use of pretreated flours, whereby starch is disorganized by precooking in a separate plant before pasta-making, is one of the processes currently used to prepare GF pasta (Marti and Pagani et al., 2013). In this regard, several heat-treatments have been proposed and each of them specifically affects starch properties. Annealing (ANN), consisting of the treatment of starch in excess of water (more than 40%) at a temperature below gelatinization, and heat-moisture treatment (HMT; treatment at low moisture and high temperatures) are hydrothermal processes often used in modifying the native physiochemical properties of starch. Both ANN and HMT increase starch crystallinity, granule rigidity, and polymer chain associations. These particular hydrothermal treatments suppress granule swelling, retard gelatinization, and increase starch paste stability, thus improving cooking behavior and texture properties (Marti and Pagani et al., 2013).

Dib and co-workers (29) evaluated the effect of hydrothermally treated CF (HTCF) addition as an improver in the manufacturing of GF laminated corn-based pasta supplemented with field bean. Treated CF was added in amounts ranging from 0 to 14.82 g for 100 g of the recipe. The mixture of hydrothermally treated corn semolina and field bean semolina (*Viciafaba minor*) was in a ratio of 2:1 (w/w) cereal-leguminous.

The optimum recipe with low CL and stickiness and higher WAC and firmness had 59.25 g of corn semolina, 7.41 g of hydrothermally treated corn semolina, 33.33 g of field bean semolina, 2 g of salt, and 77.26 mL of distilled water. The pasta produced without the addition of HTCF was considered as the control pasta. These levels of improver and water resultcd in GF pasta with good quality characteristics compared to the pasta without the improver (0 g of hydrothermally treated corn semolina). The optimum pasta had significantly lower CL (11.03%) than control pasta (16.12%). Heat treatment of CF seems to lead to the creation of less soluble components, responsible for the low CL. In addition, starch is often considered as the main structural network in GF pasta due to its functional properties. The formation of a strong network of retrograded starch around the gelatinized starch triggers less leaching of starchy components from the surface of GF pasta and, therefore, less CL.

3.4 NIXTAMALIZED CORN PRODUCTS

Currently, one of the most important industrial food products is the nixtamalized flour for making tortillas, which is obtained from grains that are nixtamalized, ground, and dried. The current versatility of this product has meant that its consumption is increasing because it is necessary only to rehydrate it to obtain dough from which the tortillas are made and therefore all of the derivatives of the same (Gomez et al., 1987).

According to Mexican Official Standard NMX-F-046-S-1980, nixtamalized CF is defined as the product obtained from the grinding of healthy, clean, and previously nixtamalized and dehydrated corn grains (*Z. mays*).

According to Mexican Official Standard NMX-F-046-S-1980, nixtamalized CF is defined as the product obtained from the grinding of healthy, clean, and previously nixtamalized and dehydrated corn grains (Z. mays). Traditional nixtamalization is the process of cooking and steeping corn in alkaline solution (1% calcium hydroxide w/w), and then washing it to produce nixtamal.

The practical advantage of using nixtamalized CF is that it must only be rehydrated with water to obtain dough, which is molded and cooked to obtain tortillas. The production of this type of flour has increased significantly in recent years (Bello-Perez et al., 2002).

The CF is a remarkable example, which is a very fine yellow flour that is obtained from the grinding of the grain and serves to give consistency to some dishes.

This type of flour is actually pregelatinized, where the grain has been subjected to a previous thermal process, giving it new characteristics and diversified uses. Some characteristics of pregelatinized flours include changes in properties such as water absorption capacity, solubility, and viscosity in an aqueous medium, among others (Amador-Rodríguez et al., 2010).

3.4.1 NIXTAMALIZATION PROCESS

Most of the corn available undergo a treatment called nixtamalization, which means "alkaline cooking" and causes important changes to the grain, which is then milled to obtain "masa" to make tortillas (INCAP, 1991).

The nixtamalization of corn (Villada et al., 2017) is a very old process developed by the Aztecs, which is still used to produce good quality tortillas and other food products. The word nixtamalization comes from the Nahuatl words nextli, or ashes of lime, and tamalli, which means cooked corn dough (García, 2004). The process of industrial nixtamalization has stimulated the study of grain quality characteristics both in breeding programs and in the industrial process of manufacturing nixtamalized corn products (Antuna-Grijalva et al., 2008).

The process of conversion of corn by nixtamalization is practiced mainly in the areas of the countries where the tortilla is consumed, especially Mexico, Guatemala, and the United States (Serna-Saldivar and Amaya-Guerra, 2008). The basis and procedure of the traditional process of nixtamalization have not changed over the centuries, regardless of whether the corn cooked in water with lime is handled by artisanal means or industrial with modern drying methods, for mass production or nixtamalized CF.

The nixtamalization consists of mixing a part of the corn with two parts of a lime solution at approximately 1%. The mixture is heated to 80 ° C for a period of 20–45 min and then left to stand overnight (Veles, 2004).

The cooking broth is called nejayote, and the resulting cooked corn is called "nixtamal." During cooking, biochemical reactions, cross-links, and molecular interactions are carried out that modify both the physicochemical, structural, and rheological characteristics of the dough and the structural and texture properties of the tortilla produced (Rodríguez Sandoval et al., 2005).

After the cooking, resting, and washing processes, the nixtamalized or nixtamal corn is obtained; this is passed through a mill to obtain the dough after the flour is dried.

The process serves to soften the corn kernel, facilitates grinding, and eliminates the pericarp; on the other hand, it has been shown that during the nixtamalization process nutrients present in the corn are lost before treatment, such as fibers (Cuellar, 2014). Also, during the cooking process, zein, which is a protein deficient in lysine and tryptophan, decreases its solubility, while other proteins (mainly albumin, globulin, and glutelin), which have higher nutritional value, increase its solubility and the availability of essential amino acids.

Other changes attributed to nixtamalization, which are reported are an increase of up to 2.8 times of the amino acid lysine, increase in the availability of tryptophan, and an increase in the ratio of isoleucine to leucine (Cuellar, 2014), and in relation to the addition of hydroxide calcium, an increase in the calcium content is reported up to 400%, compared to raw corn (FAO, 1993).

The products derived from the nixtamalization of corn with the incorporation of legumes and the purpose of evaluating the effects on the physicochemical characteristics of the formulated products will be discussed below.

3.4.2 SNACKS

Corn tortillas are the staple food of Mexico, but corn chips have become the most important salty snack food in the world (De la Parra et al., 2007).

According to the Official Mexican Standard (NOM-187-SSA1-SCFI-2002), "snack products of flour, cereals, legumes, tubers or starches are defined as snacks; as well as grains, fruits, fruits, seeds or legumes with or without skin or cuticle, tubers; nixtamalized products and pork skin, which can be fried, baked, exploited, covered, extruded or roasted; added or not with salt and other optional ingredients and food additives" (Cuellar, 2014).

The snack industry plays an important role in the health of those who consume them. The ingredients used in the elaboration must provide nutritional and sensorial characteristics suitable for the current market. Many of the snacks are perceived by the consumer as "unhealthy" food due to the high contents of fat and sodium. This is the reason why, in recent years, the botanera industry has shown some innovations in this sector, launching healthier alternatives, such as the oven, multigrain, etc. (Cuellar, 2014). Snacks have become an important part of the diets of many individuals. There is a great interest in increasing dietary fiber in foods including snacks to lead them toward a healthy approach.

The consumption of snacks globally and nationally is increasingly important. The idea is that these products, besides being tasty, are more nutritiously attractive (De la Parra et al., 2007). Having a balanced nutritional profile of calories, fat, carbohydrates, proteins, vitamins, and minerals, as well as fiber and whole grains, is part of the desired requirements in a healthy snack (Ryland et al., 2001).

Nutritious snacks can be produced by the incorporation of legumes, vegetables, or fruits in their formulation. This is why certain combinations of cereals and legumes can be very convenient from a nutritional point of view. When formulated in a mixture, an increase in the amino acid balance can be obtained; therefore, ingesting cereals and legumes together give higher quality of the proteins consumed than that obtained when they are ingested separately (Hurtado et al., 2001).

Likewise, it has been demonstrated that the possible incorporation of legume flours can have a significant impact on protein intake and dietary fiber in extruded products based on corn. This could also have an impact on the texture, expansion, and acceptability of these products due to the fact that by increasing the protein and fiber content in the food matrix the extruded products have greater hardness (Anton et al., 2009; Ozer et al., 2004).

Among ready-to-eat (RTE) products, fried products possess several interesting characteristics, including a wider consumer base, long shelf-life, high comfort, and high quality (Yadav et al., 2012).

The frying of food is one of the culinary and ancient processes that have been recorded, probably one of the technical-culinary processes that allowed to prolong the shelf life of food. Worldwide, frying is one of the methods of cooking that has greater acceptability, which is reflected in the wide range that exists in the market of fried and prefried products. Its acceptance is given not only by the flavor and texture characteristic of these foods but also by the speed of preparation (Montes et al., 2016).

Frying is commonly used for the production of commercial and home-made snacks. Fried foods are considered as a source of energy and fat. In addition to increasing the digestibility of legumes, frying reduces the moisture content of food and therefore increases shelf life; it also imparts a characteristic color, texture, and flavor to the product (Tiwary et al., 2011).

Frying is one of the most popular procedures for preparing food using cooking in oil or hot fat at elevated temperatures, where the oil acts as a transmitter of heat, producing rapid and uniform heating in the food; during this process phenomenon, loss of moisture, gain of oil, denaturation of proteins, gelatinization of starches, and substantial microstructural changes occur inside the food. Changes attributed to factors such as the initial humidity of food, the quality of the oil, the temperature of the process, and the residence time of the product in the hot oil, which if not controlled, will influence the quality of the final product (Osorio-Diaz et al., 2003).

On the other hand, fried foods enjoy increasing popularity. Their preparation is easy and fast and their tasty appearance and taste correspond to the wishes of the consumer. Frying is a complex process in which the product is subjected to a high temperature to modify the surface of the product, waterproofing it in some way, to control the loss of water from inside. In this way, it is possible to preserve many of the characteristics of the food, improving, in most cases, its taste, texture, appearance, and color. This makes it possible to obtain a more "appetizing" product, which undoubtedly contributes to the successful consumption of fried products (Valenzuela et al., 2003).

The industry that produces fried snacks is becoming bigger and more important every day. In general, the manufacture of snacks can be divided into three broad categories: (a) whole products, (b) nixtamalized products, and (c) extruded products.

3.4.2.1 NIXTAMALIZED PRODUCTS (CHIPS)

The antecedent of the corn snacks was the tortilla. Tortillas and corn chips are obtained after an alkaline thermal treatment or a process of nixtamalization of the corn kernels. Then, starting from this cooked piece, it began to "oreo" and fry, which changes its organoleptic characteristics as well as giving rise to the creation of a variety of dishes (Amador-Rodriguez, 2015).

From pre-Columbian times, the Aztecs produced the totopochtli by toasting the tortillas in a hot comal. Later, it became a common practice among the housewives to fry the leftover tortillas of the food to improve their flavor. The fried tortillas were given the name of toast when they kept their original shape and chips or tortilla chips if they were molded or sectioned into parts (Veles, 2004).

The corn tortilla chips (tortilla chips) are located in the die-cut products and are tortilla snacks. The die-cutting consists of making a uniform and thin sheet by means of two plates and rollers; then, by means of a cutter or by premolding a mass portion, triangular and circular pieces are marked, among other shapes.

The quality attributes of chips that are highly appreciated in the market include the following:

(a) free from excess oil, (b) characteristic texture, (c) thin pieces, (d) without the excess of bubbles, (e) clear and bright color, and (f) free from the smell of rancidity.

The tortilla chips contain 21%–34% oil and this content varies depending on the grain, the cooking process, the granulometry of the flour or dough, time of dehydration, and others. Opaque and oily appearance is undesirable as well as too hard to break (bite) or very brittle (Amador-Rodríguez, 2015).

The economic, cultural, and social importance of these products is evident, and the study of the process of frying tortillas chips (totopos) is relevant to design tortillas of a better sensory and nutritional quality. The tortillas chips are baked and then fried, by which they absorb less oil, become crispier, and have a greater alkaline aroma than the unbaked chips (Kawas and Moreira, 2001) (Matiacevich et al., 2012).

The traditional process of making corn snacks is relatively simple; the elimination of the nixtamalization process contributes to the elimination of the elaboration of another additional product and the standardization of the final products in the case of corn snacks.

For the formulation of masses, more than one type of flour can be used, which presents differences such as granulometry, the presence of preservation additives, nutrients, etc. The amount of water necessary to hydrate the flours is also determined; similarly, the characteristics of the final product establish the percentages of water to be added. In the case of corn chips, it needs to reduce the moisture content from the dough because through the applied thermal processes you want a reduction of water to around 2%–3% of the fried product, so that higher initial moisture in the dough required more energy to reach those percentages.

One of the main operations in the preparation of snacks is frying; this type of cooking produces foods rich in fats and often very difficult to digest, doubling or tripling their calories; thus, it has been directly related to the incidence of diseases such as obesity and overweight. For its part, another method of thermal dehydration of food is baking.

The increased preference for low-fat products coupled with continued growth in the sale of baked tortilla chips poses an additional challenge to processors of nixtamalized corn products. Totopos low in fat and with the sensory characteristics of fried chips are the dream of the snack industry, in addition to the current trend to diversify a product.

During this process, the heat transfer medium is hot air, which generates a golden crust in the food and a brittle consistency, without impregnation of fat, making it healthier (Garcia et al., 2005). High temperatures form a protective "crust" that favors the conservation of the nutrients of each food inside. This type of cooking is done in a closed environment, where the type of baking depends on the oven to be used, among which stand out the conventional oven, microwave, and infrared. The conventional or gas oven represents a great economic advantage, thanks to its low operating cost (energy efficient).

Finally, the use of thermal processes plays an important role in obtaining healthy snacks. The current trend of consumers focuses mainly on low-fat foods, which mainly leads to the use of processes such as baking. So, the biggest challenge is to improve the frying process by controlling and reducing the final fat content in fried products (Amador-Rodríguez, 2015).

3.4.2.2 SNACKS MADE WITH MIXTURES OF NIXTAMALIZED CORNFLOUR AND LEGUMES

Corn fritters have become the world's most important salty snack; however, corn proteins are considered to have low nutritional quality because they are deficient in essential amino acids such as lysine and tryptophan. Furthermore, legumes such as bean (*V. faba* L.) and chickpea (*Cicer arietinum* L.) have higher protein and essential amino acids levels. Corn proteins along with legumes complement each other to produce a protein of better quality. Hernandez-Espejo (2014) studied the preparation of corn fritters mixture with BF and CHF. This study aimed to develop CHF and BF, physicochemically characterize them, and evaluate the effect of their addition on the physicochemical and sensory properties of corn fritters. CHF and BF were elaborated and characterized by proximate analysis, grain size, differential scanning calorimetry (DSC), and color. Commercial maize flour (Macsa®), which was used for the mixtures, was similarly characterized. Three mixtures were prepared based on the protein content of each ingredient, 50% maize–50% chickpea, 50% maize–50% bean, and 50% maize–25% chickpea–25% bean. 100% maize dough and fried products were used as a control. Doughs based on blends were made adding water until standard consistency, which allows machinability (easy to handle). Doughs were characterized by adhesiveness and cohesiveness (TPA), yield, color (L^*, a^*, b^* *y* ΔE^*), and moisture content. Thereafter, doughs were laminated, cut in triangles of $6 \times 6 \times 6$ cm^3, baked (5 min and 30 s at 150 °C), and fried (30 s at 180°C–190°C) to obtain tortilla chips. Tortilla chips were characterized by proximal analysis, color (L^*, a^*, b^* *y* ΔE^*), hardness, rancidity (TBA test), expansion/contraction index, moisture content, scanning electronic microscopy, and sensorial analysis. Significant differences were observed in the doughs compared to the control with the addition of BF and CHF with regard to the moisture content, cohesiveness, yield, and color. In fried products, significant differences from control were observed in texture, rancidity index, expansion/contraction index, and color. The addition of BF and CHF to CF significantly affects the doughs and fried products; however, the sensory analysis showed that consumers have an equal acceptance of the three mixtures as for the control, thus concluding that the addition of bean and CHF to maize chips is a viable option both nutritionally and sensorially.

In another research (Campos Medina, 2017), substitutions of nixtamalized cornmeal were made by pea flour obtaining three mixtures based on cornpea (90–10, 80–20, and 70–30) and a control of 100% nixtamalized CF. The pea grains were subjected to soaking, drying, and grinding to obtain flour, and so, together with the maize, they underwent a proximal, granulometric, and color analysis. The evaluation of the flours was carried out by proximal analysis, granulometry analysis, color analysis, and DSC. With the flour mixtures described above, doughs were made, and they were tested for adhesiveness, cohesiveness, moisture, yield, and color. Afterward, the dough was rolled and cut to obtain a baked product and another fried product. The snacks were characterized by a proximal analysis, expansion index, humidity, color, texture, rancidity, and sensory analysis. The results were evaluated by means of analysis of variance (ANOVA) and compared by means with the Tukey test.

The results of this study indicate that the partial substitution of nixtamalized CF for pea flour affected most of the physical, chemical, and sensory properties, both in doughs and in baked and fried products ($p < 0.05$); however, others remained unchanged. During the characterization of the fried and baked products, changes in the composition were found, mainly with an increases in the content of protein, ash, and fiber. The baked products had a significant change with respect to the fried ones since they presented a higher percentage of proteins.

On the other hand, with the sensory evaluation, it was possible to determine the degree of satisfaction of the snacks, which was reduced according to the degree of substitution of pea flour for CF, indicating an acceptance toward the products with 10% of pea flour added. It was also determined that fried products had greater acceptability than baked products. The good acceptance of this snack has proven to be a novel and practical option, highlighting its important contribution of proteins, minerals, and fibers. Therefore, it is concluded that the use of pea flour to partially replace maize flour is a valid option to produce foods with higher nutritional quality.

Ramírez-Ortega and Rodríguez-Aguirre (2017) obtained lentil flour of a soaking process and another of germination to increase the digestibility of the components of the grain as well as the decrease of antinutritional factors. Through a series of physicochemical and mechanical tests, the appropriate germination time was determined for the preparation of the snacks.

When carrying out the corn nixtamalized flour mixtures with lentil flour and corn nixtamalized flour with germinated lentil flour, the level of substitution that can be made to baked and fried snacks was known. The process variables for the preparation of the masses and snack products were determined to obtain better characteristics and higher yield.

3.4.4.3 EXPANDED SNACKS BASED ON SQUID, CORN, AND POTATO FLOUR MIXTURES

The jumbo squid (*Dosidicus gigas*) is one of the species of commercial interest as a source of protein and has the potential for the development of RTE food products including expanded snacks by extrusion (SEE). Valenzuela-Lagarda (2017) determined the effect of the conditions of the extrusion process on the physicochemical properties of SEE made with 40% squid flour (HC) and 60% of potato (HP)/maize (HM) flour mixes in ratios (w/w) of 10:50, 30:30, and 50:10. The best values of expansion rate (2.87), bulk density (0.14 g/cm^3), firmness (7.22 N), water solubility index (27.76%), and water absorption index (5.64 g H_2O/g sample) were obtained with the mixture of higher content of potato flour. It is concluded that it is possible to produce RTE expanded snacks from squid meal rich in protein (>40%) and good physicochemical properties.

3.5 CORN BREAKFAST CEREALS

Breakfast cereals evolved innovating in elaboration and formulation processes, optimizing the nutritional value of its products. There are recent studies that point out the potential of breakfast cereals as functional foods due to the high content that many of them have of fibers and also antioxidants, nutrients of great importance in the prevention of numerous diseases.

Breakfast cereals are defined as flaked or expanded cereals made from healthy, clean, and good quality cereal grains. Their preparation may include whole grains or their parts, and in some cases, milling products (Miskelly, 2017). Breakfast cereals also called RTE cereals (Wrigley, 2010). RTE cereals are mainly made from wheat, corn, oats, rice and barley, and some mixtures of these cereals and legumes to improve their nutritional quality as a source of proteins (Jones, 2010). These products are frequently formulated from a combination of grains, legumes, and nuts and are customarily eaten with milk. These combinations help make a complete protein because the legumes, nuts, and milk provide lysine and other limiting amino acids (Serna-Saldivar, 2016). Likewise, corn grains have been used as a source of GF breakfast cereals. Recently, the extrusion of legumes has been explored to improve the nutritional quality of extruded food products; some research studies focus on the characterization of extruded soybeans, lentils, chickpeas, black beans, or pinto beans (Estrada-Girón et al., 2015).

Other ingredients include malt, salt, sweeteners (sugars), flavorings, colors, and often vitamins, minerals, and fibers. Batch processing usually involves cooking using direct steam injection followed by flaking and toasting. In continuous extrusion cooking, ingredients are mixed under heat and shear before pressing through a die to form an intermediate pellet or finished products. Popular breakfast cereals include flakes, puffed cereals, breakfast biscuits, and combinations with seeds, nuts, and dried fruits (Wrigley, 2010).

The traditional batch process may be used to produce corn flakes from a blend of maize grits, chunks of about one-half to one-third of a kernel in size, plus the addition of flavored materials, e.g., 6% (on grits wt.) of sugar, 2% of malt syrup, 2% of salt, and heat-stable vitamins and minerals. Extruded flakes, made from maize or wheat, are cooked in an extrusion cooker rather than in a batch pressure cooker and can be made from fine meal or flour rather than from coarse grits alone (Pitts et al., 2014). Corn flakes is still the most popular breakfast cereal in the world, and this product can be manufactured following the traditional process or from extrusion. Extrusion is currently the most used operation for the production of breakfast cereals and cereal-based snacks due to its versatility, faster production rate, lower investment, and lower energy (Fast, 1990).

The composition of breakfast cereals as well as the type of process used will significantly affect their functional and nutritional properties. The most strongly influenced physicochemical properties are based on the mechanical, rheological, structural, and thermal properties of breakfast cereals.

The expansion of extrudates is an important property, which is affected by the screw (configuration and speed) of the extruder and the composition (presence of protein) of the feed (Seker, 2005).

3.5.1 PHYSICOCHEMICAL PROPERTIES DUE TO EXTRUSION

Semasaka et al. (2010) elaborated an extruded breakfast cereal based on corn, millet, and soybean flour. To optimize the extrusion conditions, they modify the variables such as the temperature, rolling speed, feeding speed, and moisture content of the samples. The response variables were bulk density, water absorption and water solubility indexes, pasting properties, swelling power, color, and thermal properties. The formulation of the extruded breakfast cereal was calculated to have 17% of protein, 3% of fat, and 5% of dietary fiber. They found that the increase of temperature increased the bulk density, the water solubility, and pasting properties.

Estrada-Girón et al. (2015) evaluated the effect of extrusion conditions in extruded blends of corn and beans to obtain snack foods. They evaluated two types of beans (Peruano and black Querétaro) at different extrusion conditions. They found that temperature and moisture affected the physicochemical properties such color and breaking strength and the functional properties such as water absorption water solubility indexes and oil absorption capacity.

3.5.2 PHYSICOCHEMICAL PROPERTIES DUE TO INGREDIENTS AND ADDITIVES

Extrudate breakfast cereals contain significant concentrations of starch from corn, wheat, rice, potato, or other sources. Protein concentrations in expanded extrudates are generally lower than starch concentrations to promote expansion, crispness, and increase bulk density (Day and Swanson, 2013).

The physicochemical properties of breakfast cereals are not only affected by processing and the main composition of biopolymers such as carbohydrates and proteins but also by the addition of sugars, salts, and other additives incorporated into the mixtures, which will give result in the physicochemical properties of the final product. Salt and sugars are incorporated into extrusion breakfast cereals to provide sensorial attributes; nevertheless, the addition of these components may affect the structural, mechanical, and physicochemical behavior of RTE breakfast cereals.

Barret et al. (1995) studied the effect of the addition of sucrose on extrusion process parameters and also on the structural, mechanical, and thermal properties of corn-extruded breakfast cereals. Moisture in corn meal was also evaluated. They found that the mechanical energy is reduced at high moisture levels with the addition of sucrose. Also, no matter moisture content in corn meal is found; the addition of sucrose increased the bulk density and reduced the cell size. They concluded that the extrusion operation conditions may be adjusted depending on the sucrose content to produce extrudates with the optimal physicochemical properties. Pitts et al. (2014) studied the effect of the addition of salt and sugar to corn-wheat-extruded RTE breakfast cereals. Rheological properties such as shear and extensional viscosity were evaluated. The moisture content, expansion, and bulk density were also studied. They found an interaction between salt and sugar levels.

3.6 OTHER CORN PRODUCTS

The inclusion of cereals and legumes in the development of infant food supplements has been also investigated. Mensa-Wilmot et al. (2003)

formulated two cereal/legume-based weaning food supplements consisting of maize (*Z. mays*), cowpeas (*Vigna unguiculata*), and peanuts (*Arachis hypogea*) combined in the ratio of 43:42:15 (w/w) to form the peanut-containing blend; the soy-containing blend was formulated with maize, cowpeas, soybeans, and soybean oil in the ratio of 50:35:10:5 (w/w). Although the study was focused to evaluate the nutritional profile and not the functional profile through three different processes (cooking by extrusion, soaking, decorticating, and roasting) for the two formulated mixtures, they found differences in the content of lipids attributed to the pressure and heat of extrusion, which can cause disruption of the cells that contain lipids and therefore their release into the food matrix. These properties also affect the physicochemical properties of infant food supplements.

3.7 CONCLUSIONS

The increase in the consumption of GF products has led to the development of new products with ingredients different from wheat, oats, rye, among others. The products made from other cereals, such as corn, need to be strengthened due to their protein deficiency, which affects many of the functional characteristics of the product. The addition of legumes has contributed to improve the performance of these products due to the high protein value they present.

The characteristics such as color, rheological behavior, and textural properties are the main physicochemical properties evaluated in these products.

KEYWORDS

- **physicochemical properties**
- **corn and legumes**
- **thermal-alkaline treatment**
- **flours**
- **doughs**
- **breakfast cereals**

REFERENCES

Abdel-Aal, E.M. Functionality of starches and hydrocolloids in gluten-free foods. In Gluten-Free Food Science and Technology; Gallagher, E., Ed. Wiley-Blackwell: Oxford, UK, 2009, pp 200–224.

Abebe, A.; Chandravanshi, B.S. Levels of essential and non-essential metals in the raw seeds and processed food (roasted seeds and bread) of maize/corn (*Zea mays* L.) cultivated in selected areas of Ethiopia. Bull. Chem. Soc. Ethiop. 2017, 31(2), 185–199.

Amador-Rodriguez, K.Y. Desarrollo y evaluación alimentaria y funcional de totopos adicionados con huitlacoche, PhD Dissertation. Centro de Ciencias Básicas. Departamento de Química. Universidad Autónoma de Aguascalientes, Aguascalientes, Ags, México, 2015.

Amador-Rodríguez, K.Y.; Pérez-Cabrera L.E.; Bon Rosas, F. Desarrollo de un producto totopo a base de harina de maíz a partir de la sustitución parcial de harinas de nopal y soya. Tesis de Maestría en Ciencias Agrícolas, Pecuarias y de los Alimentos. Centro de Ciencias Agropecuarias, Universidad Autónoma de Aguascalientes, Aguascalientes, México, 2010, 145pp.

Anton, A.A.; Fulcher, R.G.; Arntfield, S.D. Physical and nutritional impact of fortification of corn starch-based extruded snacks with common bean (*Phaseolus vulgaris L.*) flour: Effects of bean addition and extrusion cooking. Food Chem., 2009, 113(4), 989–996.

Antuna Grijalva, O.; Rodríguez Herrera, S.A.; Arámbula Villa, G.; Palomo Gil, A.; Gutiérrez Arías, E.; Espinosa Banda, A.; Andrio Enríquez, E. Calidad nixtamalera y tortillera en maíces criollos de México. Rev. Fitotec. Mexicana, 2008, 31(3), 23–27.

Aravind, N.; Sissons, M.; Fellows, C. Can variation in durum wheat pasta protein and starch composition affect in vitro starch hydrolysis? Food Chem., 2011, 124, 816–821.

Barrett, A.; Kaletunç, G.; Rosenburg, S.; Breslauer, K. Effect of sucrose on the structure, mechanical strength and thermal properties of corn extrudates. Carbohydr. Polym., 1995, 26(4), 261–269.

Bello Pérez, L. A.; Osorio Díaz, P.; Agama Acevedo, E.; Núñez Santiago, C.; Paredes López, O. Propiedadesquímicas, fisicoquímicas y reológicas de masas y harinas de maíznixtamalizado. Agrociencia, 2002, 36(3), 319–328.

Burešová, I.; Kráčmar, S.; Dvořáková, P.; Středa, T. The relationship between rheological characteristics of gluten-free dough and the quality of biologically leavened bread. J. Cereal Sci., 2014, 60(2), 271–275.

Camelo-Méndez, G.A.; Tovar, J.; Bello-Pérez, L.A. Influence of blue maize flour on gluten-free pasta quality and antioxidant retention characteristics. J. Food Sci. Technol. 2018, 55(7), 2739–2748.

Campos Medina, K.J. Desarrollo de productos tipo botana con base en una mezcla de harina de maíz—harina de chícharo. Universidad Autónoma de Chihuahua, 2017

Carini, E.; Curti, E.;Minucciani, M.; Antoniazzi, F.; Vittadini, E.; Pasta. In Engineering Aspects of Cereal and Cereal-Based Products; Guiné, R.D.P.F., dos Reis Correira, P.M., Eds. CRC Press Taylor & Francis Group: Boca Raton, FL, USA, 2014; pp 211–238.

Cauvain, S.P.; Young, L.S. Fabricación de Pan. Editorial ACRIBIA S.A.: Zaragoza, España, 2002.

Cauvin, S.; Young, L. Productos de panadería, Ciencia, Tecnología y Práctica, 1st ed. Editorial ACRIBIA S.A.: Zaragoza, España, 2006.

Cole, M.E. Review: Prediction and measurement of pasta quality. Int. J. Food Sci. Technol. 1991, 26(2), 133–151.

Cuellar, M. Desarrollo, evaluación nutrimental y nutracéutica de una botana horneada a partir de harina de maíz (*Zea mays* L.) nixtamalizado y frijol común (*Phaseolus vulgaris L.*) cocido. Doctoral dissertation, 2014.

Das, A.K.; Bhattacharya, S.; Singh, V. Bioactives-retained non-glutinous noodles from nixtamalized dent and flint maize. Food Chem. 2017, 217, 125–132.

Day, L.; Swanson, B.G. Functionality of protein-fortified extrudates. Compr. Rev. Food Sci. Food Saf., 2013, 12(5), 546–564.

De la Parra, C.; Serna Saldivar, S.O.; Liu, R.H. Effect of processing on the phytochemical profiles and antioxidant activity of corn for production of masa, tortillas, and tortilla chips. J. Agric. Food Chem., 2007, 55(10), 4177–4183.

De la Peña, E.; Manthey, F.A. Ingredient composition and pasta:water cooking ratio affect cooking properties of nontraditional spaghetti. Int. J. Food Sci. Technol. 2014, 49(10), 2323–2330.

Dib, A.; Wójtowicz, A.; Benatallah, L.; Bouasla, A.; Zidoune, M.N. Effect of hydrothermal treated corn flour addition on the quality of corn-field bean gluten-free pasta. BIO Web Conf. 2018, 10, 02003.

Elías, L.G. Concepto y tecnologías para la elaboración y uso de harinas compuestas. Boletín de la Oficina Sanitaria Panamericana (OSP)., 1996, 121(2), 179–182.

Estrada-Girón, Y.; Martínez-Preciado, A.H.; Michel, C.R.; Soltero, J.F.A. Characterization of extruded blends of corn and beans (*Phaseolus vulgaris*) cultivars: Peruano and Black-Querétaro under different extrusion conditions. Int. J. Food Prop., 2015, 18(12), 2638–2651.

FAO. El maíz en la nutrición humana. Depósito de documentos de la FAO Organización de las Naciones Unidas para la Agricultura y la Alimentación: Roma,1993.

Fast, R.B.; Caldwell, E.F. Manufacturing technology of ready-to-eat cereals. In Breakfast Cereals and How They Are Made. American Association of Cereal Chemists. 1st ed.; Wiley St. Paul, MN, USA: 1990, 15–42.

Flores-Silva, P.C.; Berrios, J.D.J.; Pan, J.; Agama Acevedo, E., Monsalve González, A.; Bello-Pérez, L.A. Gluten-free spaghetti with unripe plantain, chickpea and maize: physicochemical, texture and sensory properties, CyTA – J. Food, 2014, 13(2), 159–166.

Fuad, T.; Prabhasankar, P. Role of ingredients in pasta product quality: A review on recent developments. Crit. Rev. Food Sci. Nutr. 2010, 50(8), 787–798.

Gao, Y.; Janes, M.E.; Chaiya, B.; Brennan, M.A.; Brennan, C.S.; Prinyawiwatkul, W. Gluten-free bakery and pasta products: prevalence and quality improvement. Int. J. Food Sci. Technol. 2017, 53(1), 19–32.

García, S. Estudio nutricional comparativo y evaluación biológica de tortillas de maíz elaboradas por diferentes métodos de procesamiento. Doctoral dissertation, Tesis de maestría. CICATA, Querétaro, 2004.

García, M. E. Alimentos Libres De Gluten: Un Problema Aún Sin Resolver. Redalyc (Red de revistas cientificas de america latina y el caribe, España y Portugal) 2006, 9, 123–130.

García, G.C.; Duarte, H.J.; de Hernández, L.G.; Moncada, L.M. Determinación del tiempo de cocción en los procesos de freído y horneado de tres alimentos de consumo masivo en Colombia. Épsilon, 2005, (4), 7–18.

Giménez, M.A.; González, R.J.; Wagner, J.; Torres, R.; Lobo, M.O.; Samman, N.C. Effect of extrusion conditions on physicochemical and sensorial properties of corn-broad beans (*Vicia faba*) spaghetti type pasta. Food Chem. 2013, 136, 538–545.
Giuberti, G.; Gallo, A.; Cerioli, C.; Fortunati, P.; Masoero, F. Cooking quality and starch digestibility of gluten free pasta using new bean flour. Food Chem. 2015, 175, 43–49.
Gómez, M.; Oliete, B.; Rosell, C.M.; Pando, V.; Fernández, E. Studies on cake quality made of wheat-chickpea flour blends. LWT—Food Sci. Technol., 2008, 41(9), 1701–1709.
Gómez, R.; Pérez, A.; González, A. OCDE-FAO Perspectivas Agrícolas. Universidad Autónoma Chapingo: Texcoco, México, 2013.
Gómez M.H.; Waniska R.D.; Rooney-Pflugfelder, R.L. Dry corn masa flours for tortilla and snack production. Cereal Foods World. 1987, 32, 372.
Gwirtz, J.A.; Garcia-Casal, M.N. Processing maize flour and corn meal food products. Ann. N. Y. Acad. Sci. 2013, 1312(1), 66–75.
Hernández-Espejo G.E. Evaluación de un tratamiento mecánico y extrusión-cocción en maíz para evaluar cambios físicos, químicos, propiedades reológicas, y estructurales en harina, masa, y tortilla. Tesis De Maestría en Ciencias en Ciencia y Tecnología de Alimentos. Universidad Autónoma de Chihuahua, 2014.
Hurtado, M.L.; Escobar, B.; Estévez, A.M. Mezclas legumbre/cereal por fritura profunda de maíz amarillo y de tres cultivares de frejol para consumo" snack. Arch. Latinoamer. Nutr., 2001, 5, 303–308.
INCAP. Aprovechamiento de los recursos localmente disponibles: Cereales y sus productos (Using available local resources: Cereals and their products). INCAP: Guatemala, 1991, 14 p. ilus. Lo: GT3.1,Esp/INCAP/PP/001/06
Jones, J.M. Cereal nutrition. Fast, RB; Caldwell, E. F.: Breakfast Cereals and How They Are Made. American Association of Cereal Chemists International: St. Paul, 2010, pp 411–442.
Jyotsna, R.; Milind, Sakhare, S.D.; Inamdar, A.A.; Rao, G.V. Effect of green gram semolina (Phaseolus aureus) on the rheology, nutrition, microstructure and quality characteristics of high-protein pasta. J. Food Process. Preserv., 2014, 38(4), 1965–1972.
Kaur, A.; Shevkani, K.; Singh, N.; Sharma, P.; Kaur, S. Effect of guar gum and xanthan gum on pasting and noodle-making properties of potato, corn and mung bean starches. J. Food Sci. Technol., 2015, 52(12), 8113–8121.
Kawas, M.L.; Moreira, R.G. Characterization of product quality attributes of tortilla chips during the frying process. J. Food Eng. 2001, 47(2), 97–107.
Laleg, K.; Barron, C.; Cordelle, S.; Schlich, P.; Walrand, S.; Micard, V. How the structure, nutritional and sensory attributes of pasta made from legume flour is affected by the proportion of legume protein. LWT—Food Sci. Technol., 2017, 79, 471–478.
Larrosa, V.; Lorenzo, G.; Zaritzky, N.; Califano, A. Dynamic rheological analysis of gluten-free pasta as affected by composition and cooking time. J. Food Eng., 2015, 160, 11–18.
Li, M.; Zhu, K.X.; Guo, X.-N.; Brijs, K.; Zhou, H.M. Natural additives in wheat-based pasta and noodle products: Opportunities for enhanced nutritional and functional properties. Compr. Rev. Food Sci. Food Saf., 2014, 13(4), 347–357.
Madeira Moreira da Silva, E.; Ramírez Ascheri, J.L.; Ramírez Ascheri, D.P. Quality assessment of gluten-free pasta prepared with a brown rice and corn meal blend via thermoplastic extrusion. LWT—Food Sci. Technol., 2016, 68, 698–706.
Marconi, E.; Messia, M.C. Pasta made from nontraditional raw materials: Technol nut aspects. In Durum Wheat Chemisty and Technology, 2nd ed.; Sissons, M., Abecassis, J.,

Marchylo, B., Carcea, M., Eds. American Association of Cereal Chemists International: St. Paul, MN, 2012, pp 201–211.

Marti, A.; Pagani, M.A. What can play the role of gluten in gluten free pasta? Trends Food Sci. Technol., 2013, 31(1), 63–71.

Mensa-Wilmot, Y.; Phillips, R.D.; Lee, J.; Eitenmiller, R.R. Formulation and evaluation of cereal/legume-based weaning food supplements. Plant Food Hum. Nutr., 2003, 58 (3), 1–14.

Merayo, Y.A.; González, R.J.; Drago, S.R.; Torres, R.L.; De Greef, D.M. Extrusion conditions and zea mays endosperm hardness affecting gluten-free spaghetti quality. Int. J. Food Sci. Technol., 2011, 46(11), 2321–2328.

Mirhosseini, H.; Abdul Rashid, N.F.; Tabatabaee Amid, B.; Cheong, K.W.; Kazemi, M.; Zulkurnain, M. Effect of partial replacement of corn flour with durian seed flour and pumpkin flour on cooking yield, texture properties, and sensory attributes of gluten free pasta. LWT—Food Sci. Technol., 2015, 63(1), 184–190.

Miskelly, D. Chapter 24—Optimisation of end-product quality for the consumer. In Cereal Grains (2nd edition). Assessing and Managing Quality: Woodhead Publishing Series in Food Science, Technology and Nutrition, Wrigley, C.; Batey, I.; Miskelly, D. Eds.; Elsevier: United Kingdom, 2017; pp. 653–687.

Montes, N.; Millar, I.;Provoste, R.; Martínez, N.; Fernández, D.; Morales, G.; Valenzuela, R. Absorción de aceite en alimentos fritos. Rev. Chilena Nutr., 2016, 43(1), 87–91.

Moreira, T.; Pirozi, R.; Borges, S.; Duke, U. Elaboração de pão de sal utilizando farinha mista de trigo e linhaça. Alimentos Nutrição Araraquara, 2007, 18(2), 141–150.

Morreale, F.; Boukid, F.; Carini, E.; Federici, E.; Vittadini, E.; Pellegrini N. An overview of the Italian market for 2015: cooking quality and nutritional value of gluten-free pasta. Int. J. Food Sci. Technol. 2019, 54, 780–786.

O'Shea, N.; Arendt, E.; Gallagher, E. State of the art in gluten-free research. J. Food Sci. 2014, 79(6), R1067–R1076.

Osorio-Díaz, P.; Bello-Pérez, L.A.; Sáyago-Ayerdi, S.G.; Benítez-Reyes, M.D.P.; Tovar, J.; Paredes-López, O. Effect of processing and storage time on in vitro digestibility and resistant starch content of two bean (*Phaseolus vulgaris* L) varieties. J. Sci. Food Agric., 2003, 83(12), 1283–1288.

Ozer, E.; Ibanoglu, S.; Ainsworth, P.; Yagmur, C. Expansion characteristics of a nutritious extruded snack food using response surface methodology. Eur. Food Res. Technol., 2004, 218(5), 474–479.

Padalino, L.; Mastromatteo, M.; Lecce, L.; Spinelli, S.; Conte, A.; Del Nobile, M.A. Optimization and characterization of gluten-free spaghetti enriched with chickpea flour. Int. J. Food Sci. Nutr., 2014, 66(2), 148–158.

Palavecino, P.M.; Bustos, M.C.; HeinzmannAlabí, M.B.; Nicolazzi, M.S.; Penci, M.C.; Ribotta, P.D. Effect of ingredients on the quality of gluten-free sorghum pasta. J. Food Sci. 2017, 82(9), 2085–2093.

Pérez-Hernández W.T.; De la Cruz-Magaña Y.; Miranda-Cruz E.; Ochoa-Flores A. A.; Corzo-Sosa C. A.; López-Hernández E.; Hernández-Rodríguez C.O.; Rodríguez-Blanco L. Evaluación proximal y mineral en pan de caja de harinas compuestas a base de trigo, leguminosas y oleaginosas. Investig. Desarrollo Ciencia Tecnol. Alimentos, 2018, 3, 41–47.

Pitts, K.F.;Favaro, J.; Austin, P.; Day, L. Co-effect of salt and sugar on extrusion processing, rheology, structure and fracture mechanical properties of wheat–corn blend. J. Food Eng., 2014, 127, 58–66.

Ramírez Ortega, A.; Rodríguez Aguirre, P. Elaboracion de mezclas de harinas de maiz-lenteja y maiz-germinado de lenteja para el desarrollo de productos tipo botana. Universidad Autónoma de Chihuahua, 2017.

Rodríguez Sandoval, E;, Fernández Quintero, A.; Ayala Aponte, A. Reología y textura de masas: Aplicaciones en trigo y maíz. Ing. Investig., 2005, 25(1), 72–78.

Rosenthal, A. Textura de los Alimentos: Medida y percepción. Editorial ACRIBIA S.A.: Zaragoza, España, 2001.

Roudot, A.C. Reologia y análisis de la textura de los alimentos. Editorial ACRIBIA S.A.: Zaragoza, España, 2004.

Ryland, D.; Vaisey-Genser, M.; Arntfield, S.D.; Malcolmson, L.J. Development of a nutritious acceptable snack bar using micronized flaked lentils. Food. Res. Int., 2001, 43(2), 642–649.

Saleh, M.; Al-Ismail, K.; Ajo, R. Pasta quality as impacted by the type of flour and starch and the level of egg addition. J. Texture Stud. 2016, 48(5), 370–381.

Sánchez-Hoyos D.; Peña Palacios G.A. Utilización de harinas compuestas de maíz y garbanzo adicionadas con fibra de cáscara de piña para sustitución de harina de trigo en productos de panificación. Tesis de Licenciatura. Universidad del Valle, Colombia, 2015.

Seker, M. Selected properties of native or modified maize starch/soy protein mixtures extruded at varying screw speed. J. Sci. Food Agric., 2005, 85(7), 1161–1165.

Serna-Saldivar, S.O. Cereal Grains: Properties, Processing, and Nutritional Attributes. 1st edn., CRC Press: Boca Raton, FL, USA, 2016.

Serna-Saldívar, S.O.; Amaya-Guerra, C. A. Nixtamalización del maíz a la tortilla: Aspectos nutrimentales y toxicológicos. In El Papel de la Tortilla Nixtamalizada en la Nutrición y Alimentación, Series Ingeniería. Rodríguez García, M.; Serna Saldivar, S.O.; Sánchez Senecio, F., Eds. Universidad de Querétaro: México, 2008, pp 105–151.

Shobha, D.; Vijayalakshmi, D.; Puttaramnaik; Asha, K.J. Effect of maize based composite flour noodles on functional, sensory, nutritional and storage quality. J. Food Sci. Technol. 2015, 52(12), 8032–8040.

Sicignano, A.; Di Monaco, R.; Masi, P.; Cavella, S. From raw material to dish: pasta quality step by step. J. Sci. Food Agric., 2015, 95(13), 2579–2587.

Sissons, M., Abecassis, J., Marchylo, B., Cubadda, R. Methods used to assess and predict quality of durum wheat, semolina and pasta. In Durum Wheat Chemistry and Technology, 2nd ed. Sissons, M., Abecassis, J., Marchylo, B., Carcea, M., Eds. American Association of Cereal Chemists International: St. Paul, MN, 2012, pp. 213–234.

Tazrart, K.; Zaidi, F.; Salvador, A.; Haros, C.M. Effect of broad bean (*Vicia faba*) addition on starch properties and texture of dry and fresh pasta. Food Chem. 2019, 278, 476–481.

Tiwari, U.; Gunasekaran, M.;Jaganmohan, R.;Alagusundaram, K.; Tiwari, B.K. Quality characteristic and shelf life studies of deep-fried snack prepared from rice brokens and legumes by-product. Food Bioprocess. Technol., 2011, 4(7), 1172–1178.

Turnbull, K. Quality assurance in a dry pasta factory. In Pasta and Semolina Technology. Kill, R.C. and Turnbull, K., Eds. Blackwell Science Ltd.: Oxford, UK, 2001, pp 181–221.

Umaña, J.; Álvarez, C.; Loperay, S.M.; Gallardo, C. Caracterización de harinas alternativas de origen vegetal con potencial aplicación en la formulación de alimentos libres de gluten. Rev. Alimentos Hoy, 2013, 22(29), 36–46.

Valenzuela, A.; Sanhueza, J.; Nieto, S.; Petersen, G; Tavella, M. Estudio comparativo, en fritura, de la estabilidad de diferentes aceites vegetales. Grasas y Aceites, 2003, 13(4), 568–573.

Valenzuela-Lagarda,JL.; Gutiérrez-Dorado, R.; Pacheco-Aguilar, M.E.; Lugo-Sánchez, J.B.; Valdez-Torres, C.; Reyes-Moreno, M.A.; Mazorra-Manzano; Muy-Rangel. Botanas expandidas a base de mezclas de harinas de calamar, maíz y papa: Efecto de las variables del proceso sobre propiedades fisicoquímicas, CyTA—J. Food, 2017, 15(1), 118–124.

Veles, M.J.J. Caracterización de tostadas elaboradas con maíces pigmentados y diferentes métodos de nixtamalización. Doctoral dissertation, Tesis de maestría. CICATA, Querétaro, 2004.

Villada, J.A.; Sánchez-Sinencio, F.; Zelaya-Ángel, O.; Gutiérrez-Cortez, E.; Rodríguez-García, M.E. Study of the morphological, structural, thermal, and pasting corn transformation during the traditional nixtamalization process: From corn to tortilla. J. Food Eng., 2017, 212, 242–251.

Wrigley, C. Chapter 2—The cereal grains: Providing our food, feed and fuel needs. In Cereal Grains (2nd edition) Assessing and Managing Quality: Woodhead Publishing Series in Food Science, Technology and Nutrition, Wrigley, C.; Batey, I.; Miskelly, D. Eds.; Elsevier: United Kingdom, 2017; p. 27–41.

Yadav, R.B.; Yadav, B.S.; Dhull, N. Effect of incorporation of plantain and chickpea flours on the quality characteristics of biscuits. J. Food Sci. Technol., 2012, 49(2), 207–213.

CHAPTER 4

ANALYSIS OF PHYSICOCHEMICAL AND NUTRITIONAL PROPERTIES OF RAMBUTAN (*NEPHELIUM LAPPACEUM* L.) FRUIT

JOSÉ C. DE LEÓN-MEDINA[1], CRISTIAN HERNÁNDEZ-HERNÁNDEZ[1], LEONARDO SEPÚLVEDA[1], ADRIANA C. FLORES-GALLEGOS[1], JOSÉ SANDOVAL-CORTÉS[2], JOSE J. BUENROSTRO-FIGUEROA[3], CRISTÓBAL N. AGUILAR[1], and JUAN A. ASCACIO-VALDÉS[1*]

[1]*Bioprocesses & Bioproducts Group, Food Research Department, School of Chemistry, Autonomus University of Coahuila, Saltillo 25280, Coahuila, México*

[2]*Analytical Chemistry Department, School of Chemistry, Autonomous University of Coahuila, Saltillo 25280, Coahuila, México*

[3]*Research Center for Food and Development A.C., 33088 Cd. Delicias, Chihuahua, México*

[*]*Corresponding author. E-mail: alberto_ascaciovaldes@uadec.edu.mx*

ABSTRACT

Rambutan is an exotic fruit native to Southeast Asia. It is composed of peel, pulp, and seed; the pulp is the only edible part of the fruit. The cultivars of rambutan have been expanded in many parts of the world, and its consumption is principally fresh but also is processed in the industry in the production of juices, jams, or jellies, resulting in the peel and the seed as byproducts with potential applications in the industry due to their physicochemical properties, especially the seed. The peel has been reported as an interesting source of bioactive compounds and a source of lignocellulosic material that can be used for applications such as bionanocomposites. Rambutan seed is the part of the fruit where most physicochemical properties are reported and is considered for its high amount of fat and starch. The physicochemical properties of rambutan seed fat and starch can be compared with other commercial raw materials used widely in the industry, like cocoa butter or corn starch. So, rambutan seed seems to be a potential new source to be used as a raw material in the pharmaceutical, cosmetic, and food industries.

4.1 INTRODUCTION

Rambutan is an exotic fruit that belongs to the family *Sapindaceae* and is native to Southeast Asia, particularly in the countries of Malaysia, Indonesia, and Thailand (Rohman, 2017). The name rambutan is derived from the Malayan word "rambut," which means hair; this due to the soft spines on the cortex of the fruit that resemble hair (Akhtar et al., 2018). Rambutan is a drupe with an ellipsoid or ovoid shape, and the color of the fruit varies from yellow, orange, red, and maroon (Mahmood, 2018a). The rambutan fruit is composed of 45.7% peel, 44.8% pulp, 9.5% seed, and 6.1% embryo (Solís-Fuentes et al., 2010). Another similar study showed that rambutan is comprised of the following parts: 38.6%–70.8% peel, 19.1%–45.9% pulp, and 8.3%–20.3% seed (Chai et al., 2018). The pulp of rambutan is white or translucent and generally has a sweet or mild sour flavor. The rambutan seed is covered with the pulp, and its shape is ovoid or oblong. The rambutan peel has an approximate thickness of 2–4 mm, and it is covered with fleshy pliable spinterns of 0.5–2 cm length (Figure 4.1) (Morton, 1987).

Nowadays, the fruit is consumed worldwide, principally fresh, but also is processed in the industry for the elaboration of juices, jams, marmalades, or jellies (Mahmood et al., 2018a). Because of this, waste is generated, especially of the peel and seed. Therefore, obtaining new processed products from the fruit byproducts could reduce waste and use these residues or the underutilized part for industrial purposes (Chai et al., 2018; Rohman, 2017). However, it is important to point out the importance of the physicochemical properties of the rambutan fruit to improve the applications of the rambutan fruit in the cosmetic, pharmaceutical, and food industries (Santana-Meridas et al., 2012).

So, the present work summarizes all of the physicochemical and nutritional properties that have been reported for the rambutan pulp, peel, and seed and the potential applications of these properties in the industry.

FIGURE 4.1 Rambutan fruit.

4.2 PHYSICOCHEMICAL AND NUTRITIONAL PROPERTIES OF RAMBUTAN PULP

The pulp has been characterized by a high content of total soluble solids, especially the varieties from the state of Chiapas, Mexico, which were studied for their physicochemical characteristics obtaining a fresh weight of 40.2 g, the concentration of total soluble solids of 22.62 °Brix, total sugars 423 mg 100 g^{-1}, vitamin C 37.79 mg 100 g^{-1}, and titratable acidity 0.30% (Avendaño-Arrazate et al., 2018). The water content in the fruit is the highest with 83 g, with the caloric value of 63 cal, proteins 0.8 g, carbohydrates 14.5 g, calcium 25 mg, vitamin C 20–45 mg, and iron 3 mg, all based on 100 g of pulp (Hernandez et al., 2019).

On the other hand, they also studied the macronutrients like N (77–87 mg), K (63–81 mg), Ca (22–31 mg), P (11–13 mg), Mg (9–13 mg), and S (4–6 mg), micronutrients like Mn (0.26–0.38mg), Fe (0.16–0.23 mg), B (0.12–0.16 mg), Zn (0.09–0.11 mg), and Cu (0.08–0.10 mg) in rambutan pulp (Vargas, 2003). The fresh proximal analysis presents a moisture content of 78.46, ash 0.60, crude protein 0.66, crude fiber 0.38, fat 0.24, CHO 19.66, and caloric value 83.44 K/cal, based on g/100 g (Fila et al., 2013). Also, the rambutan pulp has been analyzed for sugar content, mainly sucrose (5.39%–7.74%), fructose (2.34%–2.48%), and glucose (2.05%–2.25%); in addition, citric acid is the main organic acid in the rambutan pulp with (0.29–0.53 g/100 g) in rambutan variety R7 and wild type (Chai et al., 2017).

The rambutan pulp also presents many volatile compounds that can act as odorants. Ong et al. (1998) made extraction of the juice of rambutan obtained by the pulp using ethyl acetate and Freon as solvents and characterized the compounds by gas chromatography–mass spectrometry, obtaining 20 principal compounds that are shown in Table 4.1. The most abundant odor compounds in rambutan were cinnamic acid, 3-phenylpropionic acid, furaneol, and phenylacetic acid. On the other hand, the compounds that had the highest odor activity value were β-damascenone, ethyl 2-methylbutirate, 2,6-nonadienal, and (E)-2-nonenal. These are the compounds responsible for the odor of the rambutan fruit.

4.3 PHYSICOCHEMICAL PROPERTIES OF RAMBUTAN PEEL

The rambutan peel represents almost 50% of the total weight of the fruit, and like the peel of citrus fruits, the rambutan peel contains pectin in the cell wall, even if in less quantity, which is 1.05%–1.9% of the weight of pectin in the fresh peel. The pectin obtained from the fresh peel has a low methoxy content (11.0%–11.7%), with 43% moisture content and 8.73% ash content. The nutritional property that it contains is based on moisture (72.05), lipids (0.23), carbohydrates (23.78), protein (2.04), fiber (0.7), and ash (1.2), all of these proximal analysis values based on g/100 g (Mahmood et al., 2018b).

Also, in the rambutan peel, it has been determined the content of minerals obtained in mg/L of the dried peel of Mexican variety, such as Cu (0.070), Mn (0.14), Fe (0.29), Zn (0.080), Mg (0).15), K (0.57), Na (0.04), and Ca (0.51); these minerals provide the added value to the rambutan peel as a source of recovery along with with the content of polyphenols reported (Hernández et al., 2017). Moreover, the rambutan peel has a high lignin content of 35.34% (w/w), cellulose of 24.28% (w/w), and hemicellulose of 11.62% (w/w), which shows that besides being a potential source of bioactive compounds and possessing physicochemical compounds, it is a good source of lignocellulosic material for applications such as bionanocomposites; likewise, the application of this abundant, renewable, and low-cost waste is widespread in several areas (Oliveira et al., 2016).

TABLE 4.1 Volatile Odor Compounds Found in Rambutan Fruit (Ong et al., 1998)

Compound	μg/L of juice	Odor Activity Value
β-damascenone	2.27	226.69
(E)-4,5-epoxy-(E)-2-decenal	14.92	2.98
Vanillin	21.1	0.13
(E)-2-nonenal	7.03	87.83
Phenylacetic acid	131.67	0.02
Cinnamic acid	1340.15	0.27
Ethyl 2-methylbutyrate	15.13	151.3
δ-Decalactone	9.77	0.12
3-Phenylpropionic acid	363.09	0.02
2,6-Nonadienal	1.22	121.5
Furaneol	240.15	9.61
2-Phenyetanol	107.78	8.78
m-Cresol	5.65	0.01
Maltol	53.79	0.01
Heptanoic acid	30.43	0.08
Nonanal	51.9	62.64
Guaiacol γ-nonalactone	11.99	0.61
	29.43	35.9

4.4 PHYSICOCHEMICAL AND NUTRITIONAL PROPERTIES OF RAMBUTAN SEED

The seed of rambutan makes 4%–9% of the total weight of the fruit. It has an oblong shape, similar to that of almond, and its dimension is normally

$2.5 \times 1\ cm^2$ (Morton, 1987). Generally, the seeds are considered a byproduct in the industry of rambutan, but in some Asian countries, these are roasted and consumed (Solís-Fuentes et al., 2010).

The rambutan seed contains principally lipids, which represents approximately 37%–38%, also 11.9%–14.1% proteins, 2.8%–6.6% crude fibers, 2.6%–2.9% ash, and 28. 7% carbohydrates (Manaf et al., 2013; Augustin and Chua, 1988). Rambutan seeds contain a low amount of some antinutritional compounds like alkaloids, saponins, tannins, and phenols, which give them a bitter taste, the principal reason why theseare discarded as a byproduct (Akhtar et al., 2018). However, the high amount of lipids makes it a great source to study some physicochemical properties that can be applied for its use in the industry (Sirisompong et al., 2011). Some applications of the lipids extracted from rambutan seed have been the elaboration of soaps, candles, and fuels (Morton, 1987). In addition, there are reports that tallow obtained from rambutan can be used as a substitute for cocoa butter due to its similar physicochemical properties (Issara et al., 2014).

4.5 CHARACTERIZATION AND PHYSICOCHEMICAL PROPERTIES OF RAMBUTAN SEED FAT

The principal kinds of lipids present in the rambutan seed are the fatty acids. The proportions of unsaturated (50.9%) and saturated (49.1%) fatty acids are almost the same, in which the more abundant are oleic acid, arachidic acid, stearic acid, and palmitic acid (Manaf et al., 2013; Sirisompong et al., 2011). The extraction of the fatty acids can be reached with different techniques, principally by the use of solvents. The solvent that appears to be more effective for the extraction is the *n*-hexane, but the fatty acids also have been recovered with petroleum ether and supercritical CO_2 (Fidrianny et al., 2015; Olaniyi et al., 2013; Soeng et al., 2015). Table 4.2 shows the total fatty acids obtained from rambutan seeds by Lourith et al. (2015).

The triacylglycerols present in the rambutan seed fat have been identified by nonaqueous reverse-phase HPLC, but there were 87.29% of unknown triacylglycerols. Table 4.3 shows the percentage of triacylglycerols present in rambutan seed fat (Manaf et al., 20013).

Some physicochemical properties of the rambutan seed have been tested to characterize the material and to know what properties are important to their application in the industry, principally in the cosmetic industry. Some of the properties tested in rambutan seed fat are moisture, acid, iodine, and peroxide content, saponification, and unsaponification (Manaf et al., 2013; Lourith et al., 2015). Table 4.4 lists the reported values of these properties in different studies.

TABLE 4.2 Fatty Acids Present in the Rambutan Seed (Lourith et al., 2015)

Fatty Acids	Composition (%)
Palmitic acid (C16:0)	5.84 ± 0.12
Palmitoleic acid (C16:1)	0.86 ± 0.02
Stearic acid (C18:0)	4.54 ± 1.86
Oleic acid (C18:1)	31.08 ± 0.75
Linoleic acid (C18:2)	2.40 ± 0.07
Arachidic acid (C20:0)	28.65 ± 0.72
Behenic acid (C22:0)	3.04 ± 0.10

TABLE 4.3 Triacylglycerols Present in Rambutan Seed Fat (Manaf et al., 2013)

Triacylglycerols	Composition (%)
LLL	2.18
PLP	0.64
POL+SLL	0.39
OOO	0.91
POO	1.40
PPO	0.90
SOO	3.0
POS	1.63
PPS	0.26
AAA	1.40
Unknown	87.29

The moisture content of the fat used in the cosmetic industry needs to have ideally low values to avoid microbial contamination. The values of moisture reported of the rambutan seed fat are optimum for its application in the cosmetic industry. Also, the values of acid and iodine are suitable for their use in industrial processes. The acid value is similar to some commercial cosmetic raw materials, like avocado, coconut, and pomegranate fats (Manaf et al., 2013; Sirisimpong et al., 2011; Solis-Fuentes et al., 2010; Sonwai and Ponprachanuvut, 2012). In the case of peroxide and iodine, value allows a reduction in the oxygenation and hydrogenation degrees (Romain et al., 2013). The saponification value of rambutan seed fat is similar to those of other fats like cocoa butter, illipe butter, kokum butter, and shea butter, so there can be a chance of using rambutan seed fat as a raw material in the production of liquid and solid soaps (Sonntag, 1979; Manaf et al., 2013).

TABLE 4.4 Physicochemical Property Values of Rambutan Seed Fat

Properties	Manaf et al. (2013)	Lourith et al. (2015)
Moisture content (%)	9.60 ± 3.52	1.77 ± 0.12
Acid value (mg KOH/g fat)	–	4.35 ± 0.00
Iodine value (g I_2/100 g fat)	50.30 ± 1.24	44.17 ± 030
Peroxide value (g/g fat)	–	1.00 ± 0.00
Saponification value (mg KOH/g fat)	182.10 ± 0.16	246.73 ± 0.10
Unsaponification value (%)	0.50 ± 0.12	0.10 ± 0.00

Another important parameter for the use of fats in the food industry, especially in the chocolate industry, is the solid fat content (SFC). This factor is related to the hardness of the material, so it is preferred that the fats used in these processes have a lower SFC. The rambutan seed fat is softer at low temperatures compared to cocoa butter, but at high temperatures, the SFC of rambutan seed fat is higher. This can be due to the different components in the fats. Also, the melting points and crystallization ranges of rambutan seed fat and cocoa butter are approximately the same, with the advantage that the rambutan seed fat contains more triglycerides. So, it is demonstrated that the rambutan seed fat can be used as a substitute for the cocoa butter in the elaboration of filled chocolate (Sirisompong et al., 2011; Perez-Martinez and Parra, 2007).

4.6 PRETREATMENTS TO IMPROVE THE NUTRITIONAL AND PHYSICOCHEMICAL PROPERTIES OF RAMBUTAN SEED FAT

As mentioned above, the rambutan seed contains some antinutritional compounds that make the seed to have a bitter flavor and be discarded principally as a byproduct. But, some techniques can reduce the antinutritional compounds present in rambutan seeds. One of the most reported methods is fermentation (Olaniyi et al., 2013; Akhtar et al., 2018).

Olaniyi et al. (2013) report that after 10 days of traditional fermentation of rambutan seeds, the polyphenols content decreased from 36.5 to 25.8 mg/g, the saponin content from 1.64 to 0.977 mg/g, and the tannin content from 6.48 to 3.12 mg/g, showing a significant difference in the content of these compounds after fermentation. It was found that the hardness of the rambutan seed after fermentation increases from 1402.78 to 1525.46 N, but the value is in the acceptable range between the values of two other nuts like almond (1261 N) and walnut (1580 N). Finally, the lipid content increased by 2.65%, the pH decreased, and the color changed due to the oxidation of the compounds present in rambutan seeds.

Speaking about rambutan seed fermentation, another study confirms that the concentration of saponins decrease from 36% to 8% and the concentration of tannins decreased from 37% to 4%. The composition of fatty acids of the fermented seed rambutan also changed, being oleic (41.02%–42.38%) and linoleic (27.91%–33.10%) the most predominant, instead of arachidic acid (Chai et al., 2019a). The seeds of rambutan have also been fermented, roasted, and ground to obtain a cocoa-like powder with a similar profile of volatile compounds to that of the cocoa powder but without toxic effects against *Artemia salina* (Chai et al., 2019b).

All of these data have shown that fermentation can be used as a pretreatment to decrease the antinutritional compounds and the bitter taste of rambutan seeds, but more studies need to be conducted to confirm the safety of the process and the product.

4.7 PHYSICOCHEMICAL PROPERTIES OF RAMBUTAN SEED STARCH

Another compound present in the rambutan seed that can be extracted and exploited for its physicochemical properties is starch. Arollado et al. (2018) extracted the starch present in the rambutan seed and evaluated some physicochemical properties, shown in Table 4.5. They compared the values of rambutan seed starch with corn starch, a common commercial starch used widely in the pharmaceutical industry, and noticed that the physicochemical properties among them were very similar. So, rambutan seed starch can be considered as a potential raw material for use in the pharmaceutical industry.

4.8 PRETREATMENTS TO IMPROVE THE NUTRITIONAL AND PHYSICOCHEMICAL PROPERTIES OF RAMBUTAN SEED STARCH

However, there have been studies where rambutan seeds have been ground to a flour and then the fat is removed to improve other physicochemical properties for diverse food applications (Phanthanapratet et al., 2012). It is demonstrated that when the rambutan seed flour is defatted with supercritical CO_2, the contents of carbohydrates and protein increase, especially the amylose content, and the properties of the rambutan seed flour are similar to those of the other commercial flours like wheat flour. This demonstrates that rambutan seed flour can be used as a substitute for other commercial flours in some food industries like confectionary (Eiamwat et al., 2015).

TABLE 4.5 Physicochemical Properties of Rambutan Starch Compared to Corn Starch

Property	Rambutan Seed Starch	Corn Starch
Amylose	11.14 ± 1.79	65.37 ± 4.49
Amylopectin	87.73 ± 5.93	30.16 ± 4.04
Bulk density (g/cm^3)	0.32 ± 0.01	0.47 ± 0.02
Tapped density (g/cm^3)	0.46 ± 0.01	0.61 ± 0.01
Compressibility index (%)	29.61 ± 0.57	22.93 ± 1.95
Hausner ratio	1.42 ± 0.01	1.30 ± 0.03
Angle of repose (°)	43.64 ± 0.79	39.06 ± 1.20
Solubility (%)	5.81 ± 0.99	3.70 ± 0.89
Swelling power (g/g)	6.45 ± 0.22	6.59 ± 0.43
Viscosity (cP)	34.17 ± 3.82	331.67 ± 12.58

Another treatment that has been applied to the removal of fat from rambutan seed flour is to extract the fat with supercritical CO_2 and then use an alkaline solution. Eiamwat et al. (2016) studied and compared some physicochemical properties (Table 4.6) of the rambutan seed fat with and without the alkaline treatment. They found that the properties that increased with the alkaline treatment were the bulk density, swelling power, water absorption capacity, emulsion capacity and stability, and viscosity. Meanwhile, the turbidity, solubility, pasting temperature, and oil absorption capacity values decreased. These values can be considered for other investigations related to the extraction and evaluation of starch from rambutan seeds.

TABLE 4.6 Physicochemical Properties of Rambutan Seed Flour With and Without Alkaline Treatment (Eiamwat et al., 2016)

Property	Untreated Rambutan Seed Flour	Alkali-treated Rambutan Seed Flour
Bulk density (g/ml)	0.36 ± 0.01	0.65 ± 0.01
Swelling power(g/g)	10.64 ± 0.20	13.84 ± 0.68
Water absorption capacity (g/g)	2.56 ± 0.01	3.90 ± 0.04
Emulsion capacity (ml/ 100 ml)	47.69 ± 1.54	61.22 ± 1.94
Emulsion stability (ml/100 ml)	34.55 ± 1.38	51.79 ± 0.89
Turbidity	2.78 ± 0.01	2.11 ± 0.01
Solubility (g/100 g)	17.69 ± 0.31	13.99 ± 0.78
Oil absorption capacity (g/g)	1.41 ± 0.04	1.25 ± 0.05
Peak viscosity (cP)	1056	3055
Break down viscosity (cP)	86	647
Final viscosity (cP)	1244	4050
Pasting temperature (°C)	89	68

Finally, the removal of fat from rambutan seed flour with alkaline treatment allows the extraction of starch, with a yield of 41.3% and a consistency of fine powder with granules of oval and round shape of 5–10 µm, classified as an A-type (Eiamwat et al., 2017).

4.9 CONCLUSION

Rambutan is a fruit that has been widely studied because of its interesting properties, including the physicochemical properties. Rambutan seeds are a byproduct generated in the industry that can be exploited as a new raw material in the pharmaceutical, cosmetic, and food industries. It has been demonstrated that the components like fats or starch present in the seed have similar physicochemical properties to those of commercial ones, like cocoa butter and corn starch. Nevertheless, more studies are needed to ensure the adequate safety, quality, and organoleptic properties of the products processed in the industry.

KEYWORDS

- **rambutan**
- **physicochemical properties**
- **fat and starch**
- **pharmaceutical**
- **cosmetic**
- **food industries**

REFERENCES

Akhtar, M. T.; Ismail, S. N.; Shaari, K. Rambutan (*Nephelium lappaceum* L.). *Fruit and Vegetable Phytochemicals: Chemistry and Human Health*, 2nd edition, 2018, pp. 1227–1234.

Arollado, E. C.; Pellazar, J. M. M.; Manalo, R. A. M.; Siocson, M. P. F.; Ramirez, R. L. F. Comparison of the physicochemical and pharmacopeial properties of starches obtained from *Artocarpus odoratissimus* Blanco, *Nephelium lappaceum* L., and *Mangifera indica* L. seeds with corn starch. *Acta Medica Philippina*, 2018, *52*(4), 360.

Augustin, M. A.; Chua, B. C. Composition of rambutan seeds. *Pertanika*, 1988, *11*(2), 211–215.

Avendaño-Arrazate, C. H.; del Carmen Moreno-Pérez, E.; Martínez-Damián, M. T.; Cruz-Alvarez, O.; Vargas-Madríz, H. Postharvest quality and behavior of rambutan (*Nephelium lappaceum* L.) fruits due to the effects of agronomic practices [Calidad y comportamiento poscosecha de frutos de rambután (*Nephelium lappaceum* L.)]. *Revista Chapingo Serie Horticultura*, 2018, *24*(1), 13–26.

Chai, K. F.; Karim, R.; Mohd Adzahan, N.; Rukayadi, Y.; Mohd Ghazali, H. Physicochemical properties of two varieties of rambutan (*Nephelium lappaceum* L.) fruit. In: *Proceedings of the International Food Research Conference (IFRC 2017)*, July 25–27, 2017. Complex of the Deputy Vice Chancellor (Research and Innovation), Universiti Putra Malaysia, 2017, pp. 165–168.

Chai, K. F.; Adzahan, N. M.; Karim, R.; Rukayadi, Y.; Ghazali, H. M. Selected physicochemical properties of registered clones and wild types rambutan (*Nephelium lappaceum* L.) fruits and their potentials in food products. *Sains Malaysiana*, 2018, *47*(7), 1483–1490.

Chai, K. F.; Adzahan, N. M.; Karim, R.; Rukayadi, Y.; Ghazali, H. M. Characterization of rambutan (*Nephelium lappaceum* L.) seed fat and anti-nutrient content of the seed during the fruit fermentation: Effect of turning intervals. *LWT—Food Science and Technology*, 2019a, *103*, 199–204.

Chai, K. F.; Chang, L. S.; Adzahan, N. M.; Karim, R., Rukayadi, Y.; Ghazali, H. M. Physicochemical properties and toxicity of cocoa powder-like product from roasted seeds of fermented rambutan (*Nephelium lappaceum* L.) fruit. *Food Chemistry*, 2019b, *271*, 298–308.

Eiamwat, J.; Wanlapa, S.; Sematong, T.; Reungpatthanapong, S.; Phanthanapatet, W.; Hank-huntod, N.; Kampruengdet, S. Rambutan (*Nephelium lappaceum*) seed flour prepared by fat extraction of rambutan seeds with SC-CO_2. *Isan Journal of Pharmaceutical Sciences*, 2015, *10*, 138–146.

Eiamwat, J.; Wanlapa, S.; Kampruengdet, S. Physicochemical properties of defatted rambutan (*Nephelium lappaceum*) seed flour after alkaline treatment. *Molecules*, *21*(4), 364.

Fidrianny, I.; Fikayuniar, L.; AndInsanu M. Antioxidant activities of various seed extracts from four varieties of rambutan (*Nephelium lappaceum*) using 2, 2-diphenyl-1-picryl-hydrazyl and 2,2′-azinobis (3-ethyl-benzothiazoline-6-sulfonic acid) assays. *Asian Journal of Pharmaceutical and Clinical Research*, 2015 *8*(5), 215–219.

Fila, W. A.; Itam, E. H.; Johnson, J. T.; Odey, M. O.; Effiong, E. E.; Dasofunjo, K.; Ambo, E. E. Comparative proximate compositions of watermelon *Citrullus lanatus*, squash *Cucurbita pepo 'L* and rambutan *Nephelium lappaceum*. *International Journal of Science and Technology*, 2013, *2*(1), 81–88.

Hernández-Hernández, C.; Aguilar, C. N.; Rodríguez-Herrera, R.; Flores-Gallegos, A. C.; Morlett-Chávez, J.; Govea-Salas, M.; Ascacio-Valdés, J. A. Rambutan (*Nephelium lappaceum* L.): Nutritional and functional properties. *Trends in Food Science and Technology*, 2019, 101–120.

Hernández, C.; Ascacio-Valdés, J.; De la Garza, H.; Wong-Paz, J.; Aguilar, C. N.; Martínez-Avila, G. C.; Aguilera-Carbó, A. Polyphenolic content, in vitro antioxidant activity and chemical composition of extract from *Nephelium lappaceum* L.(Mexican rambutan) husk. *Asian Pacific Journal of Tropical Medicine*, 2017, *10*(12), 1201–1205.

Issara, U.; Zzaman, W.; Yang, T. A. Rambutan seed fat as a potential source of cocoa butter substitute in confectionary product. *International Food Research Journal*, 2014, *21*(1), 25–31.

Lourith, N.; Kanlayavattanakul, M.; Mongkonpaibool, K.; Butsaratrakool, T.; Chinmuang, T. Rambutan seed as a new promising unconventional source of specialty fat for cosmetics. *Industrial Crops and Products*, 2016, *83*, 149–154.

Mahmood, K.; Kamilah, H.; Alias, A. K.; Ariffin, F. Nutritional and therapeutic potentials of rambutan fruit (*Nephelium lappaceum* L.) and the by-products: A review. *Journal of Food Measurement and Characterization*, 2018a, 12(3), 1556–1571

Mahmood, K.; Fazilah, A.; Yang, T. A.; Sulaiman, S.; Kamilah, H. Valorization of rambutan (*Nephelium lappaceum*) by-products: Food and non-food perspectives. *International Food Research Journal*, 2018b, *25*(3), 890–902.

Manaf, Y.N.A.; Marikkar, K.L.; Ghazali, H. Physico-chemical characterisation of the fat from red-skin rambutan (*Nephellium lappaceum* L.) seed. *Journal of Oleo Science*, 2013, *62*, 335–343.

Olaniyi, L. O.; Mehhizadeh, S. Effect of traditional fermentation as a pretreatment to decrease the antinutritional properties of rambutan seed (*Nephelium lappaceum* L.). In: *Proceedings of the International Conference on Food and Agricultural Sciences (IPCBEE)*, vol. 55, October 2013.

Oliveira, E.; Santos, J.; Goncalves, A. P.; Mattedi, S.; Jose, N. Characterization of the rambutan peel fiber (*Nephelium lappaceum*) as a lignocellulosic material for technological applications. *Chemical Engineering Transactions*, 2016, *50*, 391–396.

Ong, P. K.; Acree, T. E.; Lavin, E. H. Characterization of volatiles in rambutan fruit (*Nephelium lappaceum* L.). *Journal of Agricultural and Food Chemistry*, 1998, *46*(2), 611–615.

Pérez Martínez, V. T.; Lorenzo Parra, Z. El impacto del déficit mental en el ámbito familiar. *Revista Cubana de Medicina General Integral*, 2007, *23*(3), 1–6.

Phanthanapratet, W.; Trangwacharakul, S.; Wanlapa, S. Application of rambutan seed flour for low calorie salad dressing product. *Agricultural Science Journal*, 2012, *43*(2), 517–520.

Rohman, A. Physico-chemical properties and biological activities of rambutan (*Nephelium lappaceum* L.) fruit. *Research Journal of Phytochemistry*, 2017, *11*(2), 66–73.

Romain, V.; Ngakegni-Limbili, A. C.; Mouloungui, Z.; Ouamba, J. M. Thermal properties of monoglycerides from *Nephelium lappaceum* L. oil, as a natural source of saturated and monounsaturated fatty acids. *Industrial & Engineering Chemistry Research*, 2013, *52*(39), 14089–14098.

Santana-Méridas, O.; González-Coloma, A.; Sánchez-Vioque, R. Agricultural residues as a source of bioactive natural products. *Phytochemistry Reviews*, 2012, *11*(4), 447–466.

Sirisompong, W.; Jirapakkul, W.; Klinkesorn, U. Response surface optimization and characteristics of rambutan (*Nephelium lappaceum* L.) kernel fat by hexane extraction. *LWT – Food Science and Technology*, 2011, *44*(9), 1946–1951.

Soeng, S.; Evacuasiany, E.; Widowati, W.; Fauziah, N., Manik, V. T.; Maesaroh, M. Inhibitory potential of rambutan seeds extract and frac-tions on adipogenesis in 3T3-L1 cell line. *Journal of Experimental and Integrative Medicine*, 2015, *5*, 55–60.

Solís-Fuentes, J. A.; Camey-Ortíz, G.; del Rosario Hernández-Medel, M.; Pérez-Mendoza, F.; Durán-de-Bazúa, C. Composition, phase behavior and thermal stability of natural edible fat from rambutan (*Nephelium lappaceum* L.) seed. *Bioresource Technology*, 2010, *101*(2), 799–803.

Sonntag, N. O. V. Bailey's Industrial Oil and Fat Products. Swern, D. (Ed.), 4th ed., 1979, pp. 322–328.

Sonwai, S.; Ponprachanuvut, P. Characterization of physicochemical and thermal properties and crystallization behavior of krabok (*Irvingia malayana*) and rambutan seed fats. *Journal of Oleo Science*, 2012, *61*(12), 671–679.

Vargas-Calvo, A. Descripción morfológica y nutricional del fruto de rambután (*Nephelium lappaceum*). *Agronomía Mesoamericana*, 2003, 14(2), 201–206.

CHAPTER 5

DEVELOPMENT OF MODERN ELECTROANALYTICAL TECHNIQUES BASED ON ELECTROCHEMICAL SENSORS AND BIOSENSORS TO QUANTIFY SUBSTANCES OF INTEREST IN FOOD SCIENCE AND TECHNOLOGY

MARÍA A. ZON[1], FERNANDO J. ARÉVALO[1], ADRIAN M. GRANERO[1], SEBASTIÁN N. ROBLEDO[2], GASTÓN D. PIERINI[1], WALTER I. RIBERI[1], JIMENA C. LÓPEZ[1], and HÉCTOR FERNÁNDEZ[1*]

[1]*Grupo de Electroanalítica (GEANA), Departamento de Química, Facultad de Ciencias Exactas, Físico-Químicas y Naturales, Universidad Nacional de Río Cuarto, Agencia Postal No. 3, 5800 Río Cuarto, Argentina*

[2]*Departamento de Tecnología Química, Facultad de Ingeniería, Universidad Nacional de Río Cuarto, Agencia Postal No. 3, 5800 Río Cuarto, Argentina*

[*]*Corresponding author. E-mail: hfernandez@exa.unrc.edu.ar*

ABSTRACT

The importance of electroanalytical techniques in the determination of many substances in the science and technology of food has increased significantly in recent years. This has been possible thanks to the great advances made in the fields of electronics, biotechnology, the development of new nanomaterials, etc. They have allowed, in turn, notable advances in the development of the modern electroanalytical techniques. Some of the advantages of these techniques compared to the conventional analytical techniques are higher speed, greater simplicity, lower cost, lower consumption of solvents, and the possibility of making determinations in real samples without any pretreatment. This chapter discusses the development of electrochemical sensors and biosensors for the quantification of important components of foods (natural and from contamination), such as mycotoxins, synthetic and natural antioxidants, metals, etc., performed by the members of the GEANA** group during the last 20 years.

5.1 INTRODUCTION

In the fascinating world of food for human life, the control of food quality plays a role of significant importance. This is so given that the care of the health of the people is crucial and must be an inalienable objective for the authorities and responsible people, from the moment of production to the final consumption and during the multiple intermediate stages (Mustafa and Andreescu, 2018). Undoubtedly, the processes of quality control of food have been subject to regulations and techniques of different nature available over time (Alahi an Mukhpadhyay, 2017). There is no doubt that, based on the discoveries and scientific and technological progresses that have gradually been achieved, they have resulted in improvements in the extension and quality of life of at least that part of the world population that has had the greatest access to these advances (Segneanu et al., 2018). The growing demand for techniques of analysis of the various components of food allowed increasing the interest of the scientific and technological community in both refining existing techniques and developing new techniques and methodologies based on the results obtained from basic research, which have allowed to expand the horizons in terms of substances to be determined (Gruemezescu, 2017). On the other hand, in addition to currently having techniques to analyze an important variety of substances present in food, the challenge has also been put into developing techniques that allow the simplest determinations to be made, in increasingly shorter times, at lower cost, and, in many cases, with systematically lower detection limits (Rhazi et al., 2018). Some of these achievements have been satisfied with the development of some modern electrochemical techniques (Bard and Faulkner, 2001). These techniques begin to be applied in relatively recent times (in comparison with conventional techniques) from the tremendous progress achieved in different technological disciplines, related to the advancement in electronics, biological systems, informatics technology, new nanomaterials, etc. (Fernandez et al., 2017). The application of select electrochemical techniques to the determination of substances in the most varied real matrices has given rise to what is known as "electroanalytical chemistry," which currently has a place of significant importance within modern analysis techniques (Fernandez and Zon, 2002). Within a broad spectrum of real samples of different nature in which electrochemical techniques are used, food is receiving increasing attention (Zeng et al., 2018). Different substances present in food have been the subject of attention for the development of electroanalytical techniques for their quantification (Viswanathan et al., 2009). This is how the potential pulse voltammetric techniques (Osteryoung, 1983), stripping (Alghamdi, 2010), and electrochemical sensors and biosensors (Pividory and Alegret, 2010) emerge as good candidates. This chapter shows the contributions made in the last 20 years by the GEANA group of the National University of Rio Cuarto, Argentina, oriented to the development of electroanalytical methodologies for the determination of mycotoxins, synthetic and natural antioxidants, and other important analytes in foods. This chapter highlights the importance of combining the results of basic research, from the point of view of chemical physics, with the advantageous application of such results to the development of electroanalytical methodologies in quality control systems for food and other important areas.

5.2 ELECTROANALYTICAL TECHNIQUES

5.2.1 POTENTIAL PULSE VOLTAMMETRIES

With the development of potential pulse voltammetric techniques starts the application of electrochemical techniques for the determination of substances at trace levels, which is very important in different areas (Borman, 1982). Particularly, these techniques are the normal pulse, differential pulse, and square wave voltammetries (Osteryoung, 1983). Given the remarkable technological progress in digital and electronic techniques in instrumental design, the acquisition of low-cost electrochemical equipment for laboratory and field determinations is now very accessible, with access to the most modern electroanalytical techniques. Currently, differential pulse and square wave voltammetries are the most used, given their capabilities to provide detection limits at submicromolar levels (Zon et al., 2014).

5.2.2 *STRIPPING VOLTAMMETRY*

In addition to the advantages of the previously mentioned techniques, the preconcentration of the analyte (on or in) the electrode material has provided the possibility of quantifying substances at subnanomolar levels (Wang, 2006). This technique is currently one of the most competent ones for the determination of analytes at these levels, with low cost equipment and in reasonable times (Wang, 2006).

5.2.3 *ULTRAMICROELECTRODES*

Ultramicroelectrodes (UMEs) are electrodes where at least one of their dimensions are micrometric and whose electrochemical response differs in usual conditions from responses of conventional electrodes used in regular electrochemical experiments (Fleischmann et al., 1987). Their applications have covered a wide range of demands, from the determination of electrode kinetics, physicochemical properties in highly resistive media, etc. to analytical applications for the quantification of substances of importance in food, health, and others areas (Queiros et al., 1990).

5.2.4 *ELECTROCHEMICAL BIOSENSORS*

Within biosensors, in a wide concept (Mustafa et al., 2017), electrochemical biosensors are classified as amperometric, potentiometric, impedimetric, etc., according to the electrochemical measurement technique (Wang, 2006). Biosensors are devices that relate the interaction of a given analyte with biological material (antibody, enzyme, etc.), and, as result, an electrical signal is obtained, which, in some way, is proportional to the concentration of the analyte in the sample (Wang, 2006). Due to their appearance and the development of new nanostructured materials, these devices have shown tremendous growth in terms of the variety of substances to be determined in food matrices and other materials of importance (Fernandez et al., 2012; Mishra et al., 2018).

5.3 FOODS AND ANALYTES

Of the most variety of analytes that are present in food and, currently, are subject to quality control, GEANA has been interested in studying the electrochemical properties of mycotoxins and synthetic and natural antioxidants, among others. These substances may be present in a variety of foods, both in their version of raw material and also processed (Zon et al., 2014; Fernandez et al. 2012a, 2012b). In addition, the results of these basic studies have been useful, not only for the purpose of contributing to the training of specialized professionals but also as a base of inestimable importance for the development of electroanalytical techniques for the determination of these substances in real food samples.

5.3.1 MYCOTOXINS

Mycotoxins are secondary metabolites produced by fungi of different genera that can contaminate food at any stage of the food chain. Their effects on human health are diverse and have been widely characterized (Soriano del Castillo, 2007). Official organizations from different countries have regulated the maximum levels allowed in a wide variety of foods (IARC, 2006). Different techniques are used for their quantification (Vettorazzi et al., 2014; Fernandez, 2013). The mycotoxins studied by GEANA are (a) alternariol, alternariol monomethyl ether, altertoxin I, and altenuene, all of the *Alternaria* genus; (b) zearalenone, deoxynivalenol, and moniliformin, of the *Fusarium* genus; (c) cercosporin, of the *Cercospora* genus; and (d) citrinin, ochratoxin A, sterigmatocystin, and patulin, of the *Penicillium* and *Aspergillus* genera (Fernandez et al., 2012b).

5.3.2 SYNTHETIC ANTIOXIDANTS

Synthetic antioxidants, particularly phenolic antioxidants, have been widely used to retard oxidative rancidity in edible vegetable oils, margarins, and many other foods. However, there is currently a tendency to use natural antioxidants to avoid possible toxicity risks (Zon et al., 1999). The most common synthetic antioxidants are butylated hydroxyanisole (BHA), butylated hydroxytoluene (BHT), propyl gallate (PG), *tert*-butyl hydroxyquinone (TBHQ), etc. (Robledo et al., 2014). Their electrochemical reaction mechanisms have been opportunely proposed (Pierini et al. 2017).

5.3.3 *NATURAL ANTIOXIDANTS*

Natural antioxidants belong to different chemical families. Most of those studied by GEANA are tocopherols (α, β, δ, γ) (Robledo, 2014), several from the flavonoid family (morin (MO), rutin (RUT), fisetin (FIS), butein (BUT), luteolin (LUT)), eugenol (EUG) (Pokorny et al., 2001), and trans-resveratrol (t-RES) (Vitaglione, 2005). Tocopherols are found, among other foodstuffs, in edible vegetable oils (2014). t-RES is found mainly in red wines (Vitaglione et al., 2005), and flavonoids and EUG are found in a variety of foods (Pokorny et al., 2001). In the same way as for synthetic antioxidants, their electrochemical reaction mechanisms have been studied (Robledo et al., 2014).

5.3.4 *OTHER ANALYTES*

Metals such as lead in propolis, the herbicides molinate and atrazine in drinking water, monoterpene in essential oils, and hypoxanthine (Hx), xanthine (Xa), and uric acid in fish were also considered of interest for the food system by GEANA (Pierini et al., 2013, 2018; Arevalo et al., 2012; Gonzalez-Techera et al., 2015; Robledo et al., 2019).

5.4 ELECTROANALYTICAL METHODOLOGIES DEVELOPED BY GEANA

5.4.1 *MYCOTOXINS*

5.4.1.1 ALTERNARIA GENUS: ALTERNARIOL (AOH), ALTERNARIOL MONOMETHYL ETHER (AME), ALTERTOXINI (ATX-I) AND ALTENUENE (ALT)

Fungi of the *Alternaria* genus are widely distributed in the soil and decaying organic matter (Soriano del Castillo, 2007). Through spectroscopic and electrochemical measurements, it was demonstrated for the first time that AOH and AME are substrates of the mushroom tyrosinase enzyme, which allowed evaluating their kinetic parameters and their analytical performance through the design of an enzymatic electrochemical biosensor (Moressi et al., 1999). The electrochemical properties of ALT and ATX-I were studied by cyclic voltammetry (CV) on glassy carbon and carbon paste electrodes, respectively. ATX-I showed surface adsorption properties that allowed measurements of adsorptive stripping voltammetry. Excellent calibration curves and detection limits of 4.0×10^{-7} M for ALT and 3×10^{-9} M for ATX-I were obtained by square wave voltammetry (SWV) (Molina et al., 2000, 2002). On the other hand, AME and ATX-I showed surface properties on gold electrodes modified with self-assembled monolayers, which allowed achieving good detection limits for their determination (Moressi et al., 2004, 2007).

5.4.1.2 FUSARIUM GENUS: ZEARALENONE (ZEA), DEOXINIVALENOL (DON), AND MONILIFORMIN (MON)

Fusarium fungi produce an important variety of mycotoxins (Soriano del Castillo, 2007). Particularly, ZEA, DON, and MON have been studied by GEANA.

ZEA is an estrogenic mycotoxin that is produced by several species of *Fusarium* and is found particularly in cereals (Soriano del Castillo, 2007). Molina et al. have determined its electrochemical properties. These authors found that ZEA is adsorbed on the surface of carbon paste electrodes and used this property to develop an electrochemical method based on SWV for its determination in rice and corn samples, with detection limits on the order of 30 ppb (Molina et al, 2003; Ramirez et al., 2005). Recently, Riberi et al. (2018) developed a very sensitive amperometric electrochemical immunosensor for the determination of ZEA in corn samples. The immunoelectrode is based on carbon screen-printed electrodes modified with dispersions of multiwalled carbon nanotubes/polyethyleneimine and uses anti-ZEA polyclonal antibodies linked to gold nanoparticles immobilized on the nanotubes. A competitive assay between ZEA present in the sample and ZEA labeled with horseradish peroxidase (HRP) enzyme was performed. The detection limit found was 0.15 ppb. One of the advantages of the test lies in the use of samples without pretreatment (Riberi et al., 2018).

DON is a mycotoxin that belongs to the group of trichothecenes and is produced by several *Fusarium* species (Soriano del Castillo, 2007). It is a toxic contaminant found in a wide variety of food for animals and humans. A simple method for the indirect determination of DON based on the homogeneous reduction of the toxin byproducts of the electrochemical reduction of dissolved oxygen in solution has been described. The methodology was based on a SWV anodic adsorptive stripping procedure, and DON was determined in samples of soybean meal (Molina et al., 2008).

MON is a mycotoxin that belongs mainly to the species *Fusarium proliferatum*. MON has been found in a variety of cereals, such as rice, corn, wheat, oats, etc. However, it has been reported that the greatest contamination has been found in corn samples. The first data on the electrochemical properties of MON have been reported by researchers of GEANA (Diaz Toro et al., 2015). A very sensitive electrochemical sensor for the determination of MON in corn samples has recently been developed by GEANA. It is based on gold electrodes modified with cysteamine self-assembled monolayers. The standard additions method was used, and a detection limit of 0.1 ppb was obtained (Diaz Toro et al., 2016).

5.4.1.3 CERCOSPORAGENUS: CERCOSPORIN (CER)

The *Cercospora* genus belongs to a large group of toxin-producing pathogenic fungi that infect a wide variety of leaves in crops of great economic importance. CER is one of the nonspecific toxins produced by *Cercospora* (Daub et al., 1982). GEANA was the first to characterize CER using the conventional and pulse-potential electrochemical techniques. The detection limits of 1.9×10^{-6} M were obtained by measurements of the diffusional reduction peak current signal, while 3.7×10^{-8} M was obtained by adsorptive stripping SWV (Zon et al., 1999; Marchiando et al., 2003). The CER adsorption property on glassy carbon electrode surfaces was used to develop an electrochemical sensor for the determination of CER in leaves of naturally infected peanut obtained from the Río Cuarto region, Argentina. A detection limit of 6 ppb was determined, and good percentages of recovery were obtained on infected extracts (Marchiando et al., 2005).

5.4.1.4 PENICILLIUM AND ASPERGILLUS GENERA: OCHRATOXIN A (OTA), CITRININ (CIT), PATULIN (PAT), AND STERIGMATOCYSTIN (STEH)

Fungi of *Penicillium* and *Aspergillus* genera produce an important variety of mycotoxins widely distributed in many food sources. OTA, CIT, PAT, and STE are mycotoxins produced by these fungi, which have been studied by GEANA. OTA is a secondary metabolite produced by *Penicillium* and *Aspergillus* fungi. It has been reported to have important consequences on human and animal health as a possible carcinogen. The electrochemical properties of OTA obtained by GEANA have driven the use of analytical methodologies for its quantification (Ramirez et al., 2010). Thus, the development of a sensor based on gold electrodes modified by cysteamine self-assembled monolayers using SWV as an electrochemical technique for the quantification of OTA in red wine samples has been reported. An extremely low detection limit was reported, that is, 0.004 $\mu g\ L^{-1}$ (Perrota et al., 2011). At the same time, the affinity between OTA and the *Brassica napus* root peroxidase enzyme was found, allowing the development of an amperometric enzymatic biosensor to determine OTA in peanut samples (Ramirez et al., 2011). On the other hand, the generation of nanomaterials allowed the research and development of a highly sensitive electrochemical magnet immunosensor to determine OTA in red wine samples (Perrota et al., 2012). This device was based on the use of magnetic particles functionalized with the G protein as the solid phase to produce the affinity reaction between OTA and the monoclonal antibody of OTA immobilized on magnetic beads. The trial consisted of a direct competition between OTA present in wine samples and OTA labeled with HRP. The immunoassay was performed on carbon screen-printed electrodes as the electrochemical transduction element, and the measurements were carried out by SWV. A very low detection limit of 0.008 ppb was determined. The design of this immunosensor allows determining OTA without a previous treatment of the sample. In addition, the methodology is fast and uses small sample volumes (Perrota et al., 2012).

CIT has been found to contaminate cereals, such as wheat, barley, rice, corn, etc., among others (Arévalo et al., 2011). Although some studies reveal the possibility that CIT has toxic, teratogenic, mutagenic, and carcinogenic properties, there is no certain evidence about its toxicity in humans and animals (IARC, 2006). Zachetti et al. (2012) reported the first data about the

CIT electrochemical properties. In the same way, the first electrochemical biosensors for the quantification of CIT were developed by GEANA. Thus, an electrochemical immunosensor incorporated in a microfluid cell was developed for the determination of CIT in rice samples. Thus, it was developed a system in which CIT was immobilized on a cysteamine monolayer adsorbed on a gold film, which, in turn, was previously electrodeposited on a glassy carbon electrode. Thus, immobilized CIT competes with CIT present in the solution for the anti-CIT IgG antibody present also in solution. The method has proven to be very sensitive and selective. The detection limit obtained was 0.1 ng mL^{-1} and allowed the analysis to be performed in approximately 45 min with a minimum pre-treatment of the sample (Arévalo et al., 2011). In addition, an enzymatic amperometric biosensor was proposed as another alternative for the determination of CIT in rice samples. It consisted of an electrode obtained from a mixture of carbon paste with multiwalled carbon nanotubes, HRP, and ferrocene (Fc) as a redox mediator. A reasonable detection limit of 0.25 nM was obtained (Zachetti, 2013).

PAT is a mycotoxin that mainly contaminates apples and their derived products, although it has also been found in other fruits. The electrochemical behavior of PAT and 5-hydroxymethylfurfural, a common interfering in the determination of PAT in apple-derived products, has also been studied by GEANA (Chanique et al., 2013). Chanique et al. (2013) developed an electrochemical sensor for the determination of PAT in apple juices in the presence of 5-hydroxymethylfurfural based on SWV, with a detection limit of 45 ppb, which enables it to be used in screening tests. On the other hand, Riberi et al. (2018) reported for the first time the development of an electrochemical impedimetric immunosensor for the direct noncompetitive determination of PAT in commercial apple juices. Glassy carbon electrodes were modified with graphene oxide on which polyclonal antibodies specific for PAT were covalently anchored and a solution of ferricyanide/ferrocyanide was used as a redox probe. An important advantage of this methodology is that it is not necessary to label neither the antigen nor the antibody. A very low detection limit of 9.8 pg mL^{-1} was obtained, and the determination of PAT in juice sample was performed without previous treatment (Riberi et al., 2020).

STEH is produced not only by fungi of the *Aspergillus versicolor* genus but also by other fungal species. Since STEH is a precursor of aflatoxin B_1 (AFB_1), the acute toxicity, carcinogenicity, and metabolism of STEH are compared to those of AFB_1 and other hepatotoxic mycotoxins (Diaz

Nieto et al., 2016, 2018). The first data on the thermodynamic, kinetic, and analytical properties of STEH from electrochemical data were also published by GEANA (Diaz Nieto, 2016). A third-generation amperometric enzymatic biosensor was recently developed for the determination of STEH in corn samples. The platform used consisted of glassy carbon electrodes modified with a mixture of soybean peroxidase enzyme (SPE) and chemically reduced graphene oxide. The surface response methodology of the experimental design was used to achieve optimal working conditions. A good detection limit of 2.3×10^{-9} mol L^{-1} was obtained for a signal-to-noise ratio of 3:1 (Diaz Nieto et al., 2019).

5.4.2 SYNTHETIC ANTIOXIDANTS

Experiments performed with UME as electrochemical sensors for synthetic phenolic antioxidants determination were the pioneer works of GEANA for analytical technique development. The antioxidants studied were BHA, BHT, PG, and TBHQ (Robledo et al., 2014).

Ceballos and Fernández (2000a, 2000b) were the first to propose the use of a sensor based on UME for the determination of BHA, BHT, and PG in edible vegetable oils and fats by means of electroanalytical techniques. These authors showed the advantages of performing voltammetric measurements on UME of disk carbon fiber in two-electrode nonconventional electrochemical cells for the determination of BHA and BHT in sunflower oil and PG in pig fat (Ceballos and Fernandez, 2000a). On the other hand, they included the SWV as an electrochemical technique for the quantification of BHA and BHT in corn oil using benzene/ethanol/H_2SO_4 solutions for direct measurements and acetonitrile after an extraction process. It was demonstrated that the measurements with the proposed sensor greatly improved the analysis times compared to the conventional techniques (Ceballos et al., 2006).

A novel methodology for the qualitative and quantitative analyses of BHA, BHT, TBHQ, and PG in edible olive oils has been reported by Robledo et al. (2011). It is based on the acid–base properties of zantioxidants and the use of SWV on platinum band UME and the method of standard additions. Both disk carbon fiber and platinum band UME were successfully used to resolve mixtures of BHT and tocopherols by SWV in olive and corn oils (Robledo et al., 2014).

5.4.3 NATURAL ANTIOXIDANTS

Electrochemical sensors for natural antioxidants present in some foods were also developed by GEANA, that is, t-RES, polyphenols, MO, RUT, FIS, BUT, EUG, LUT, and tocopherols. The methods developed were mainly based on potential pulse techniques and enzymatic electrochemical biosensors.

The first analyte studied for its growing interest in the benefits of human health was t-RES (Granero et al., 2013). Granero et al. (2013) developed an amperometric biosensor based on carbon paste electrodes modified with the basic peroxidase of *Brassica napus* hairy roots (PBHR) and Fc. The oxidation of t-RES catalyzed by this isoenzyme was reported for the first time. The lowest t-RES concentration reported was 0.83×10^{-6} M (Granero et al., 2008). This biosensor was slightly modified by the addition of multiwalled carbon nanotubes to the carbon paste/PBHR/Fc composite for the determination of the total content of polyphenolic compounds in wine and tea samples (Granero et al., 2010).

The electrochemical properties of the antioxidants MO, RUT, FIS, BUT, and EUG were obtained by the CV and SWV techniques. The SWV was also used to generate analytical data, such as the calibration curves and detection limits for these antioxidants in simulated media using both conventional and modified electrodes. The lowest concentrations measured were in the range $(1–5) \times 10^{-8}$ M (3–14 ppb) (Tesio et al., 2011, 2013; Aragáo-Catunda Jr. et al., 2011; Maza et al., 2012, 2017; Lopez et al., 2015). However, the catalytic oxidation of EUG by the SPE enzyme was also studied. It was based on the immobilization of a conjugate formed by SPE and adamantane (SPE-ADA) on a glassy carbon electrode modified with a composite of chemically reduced graphene oxide and β-cyclodextrin (CRGO-βCD). Then, gold nanoparticles were generated on the CRGO-βCD/SPE-ADA electrode to improve the heterogeneous charge transfer rate. An experimental design was used to develop an enzymatic amperometric biosensor for the quantification of EUG (Lopez et al., 2019, 2020). The results obtained are auspicious and encouraging for the design of electrochemical biosensors for the determination of these antioxidants in a wide variety of foods in the future.

A highly sensitive electrochemical sensor based on glassy carbon electrodes modified with multiwalled carbon nanotubes dispersed in polyethyleneimine was developed for the quantification of LUT in samples of peanut hull. The electrode was stable for 23 days, and a very low detection limit of 5.0×10^{-10} M was obtained (Tesio et al., 2014, 2015).

On the other hand, tocopherols naturally present in a wide variety of foods have also been studied by GEANA, considering the importance they have because of their antioxidant properties. Robledo et al. (2013) have reported the development of an electrochemical sensor for the determination of tocopherol isomers (α, γ, δ) in edible vegetable oils (canola, sunflower, corn, soybean, and grape seeds oils). The technique is based on the use of disk carbon fiber UMEs. The overlap of the voltammetric signals was solved through chemometric tools that allowed obtaining a multivariate calibration model. The method of artificial neural networks was the most efficient for the quantification of tocopherols in edible vegetable oils (Robledo et al., 2013). On the other hand, mixtures of tocopherols (α, γ, δ) with a synthetic antioxidant, BHT, present in olive and corn edible oils, could be favorably resolved by extractive methods using SWV on a disk carbon fiber UME for the tocopherols in a medium of benzene and sulfuric acid and a platinum band UME in a nonaqueous medium for BHT (Robledo et al., 2014).

5.4.4 OTHER SUBSTANCES

Sensors and electrochemical biosensors for the determination of other important analytes in the food system were also developed by GEANA.

The presence of lead as a contaminant in propolis has been considered a concern for a long time. Pierini et al. developed an electrochemical sensor based on bismuth films deposited on glassy carbon electrodes. The use of bismuth, which has low toxicity, has been proposed as an alternative to the use of mercury. The detection limit was 0.6×10^{-6} g L^{-1}. It was successfully tested in the quantification of lead in contaminated samples of crude *Argentine propolis* and has been considered as a good alternative for the quality control of these products (Pierini et al., 2013).

Molinate and atrazine herbicides, which contaminate river water, have also been studied by GEANA. Arévalo et al. have developed a highly sensitive electrochemical magnet immunosensor with a wide analytical range for the determination of molinate. As a novel detail, the use of phage particles containing mimetic peptides of the analyte to be determined to substitute conventional markers was introduced. The analyte is detected when it competes with a tracer compound for the antibody. The immunosensor assay was faster than the equivalent enzyme-linked immunosorbent assay

(ELISA) assay and showed a significantly lower detection limit (Arévalo et al., 2012). An interesting variant to the previously described one was reported for the development of a new electrochemical noncompetitive phage anti-immunocomplex immunosensor for the quantification of atrazine herbicide in river water without the need of pretreatment of the sample. It was used recombinant M13 phage particles containing a peptide that can recognize the immunocomplex of atrazine with an antiatrazine monoclonal antibody. In the same way, this immunoelectrode showed better analytical performances than conventional ELISA, i.e., a significantly lower detection limit (0.2 pg mL^{-1}) and a greater range of work (González-Techera et al., 2015). In addition, it has been indicated that phages can be simply selected from phage libraries and that their production is simple and inexpensive, which encourages their use in the design of high-sensitivity biosensors (Arévalo et al., 2012; González-Techera et al., 2015).

Pierini et al. (2018) have proposed an electrochemical sensor for the simultaneous quantification of Hx, Xa, and uric acid in untreated fish samples. An edge plane pyrolytic graphite electrode was used, and SWV was the electrochemical technique. The parameters of the experiments were optimized by experimental design and the pretreatment of the electrodes was optimized by multiresponse assays. Detection limits in the range $(0.03–0.08) \times 10^{-6}$ mol L^{-1} were obtained. The method developed is simple and fast and is proposed as an excellent option for quality control of fish.

Robledo et al. (2019) developed an electrochemical sensor to quantify phenolic monoterpenes such as carvacrol and thymol in oregano and thyme essential oils. Linear sweep voltammetry on glassy carbon electrodes was the electrochemical technique used. An interesting fact of the method is that neither the sample needs a pretreatment nor the electrode surface needs modification. Thus, the proposed technique emerges as a very good alternative for quality control of essential oils.

KEYWORDS

- **modern eletroanalytical techniques**
- **sensors**
- **biosensors**
- **food science and technology**

REFERENCES

Alahi, M. E. E.; Mukhopadhyay, S. C. Detection methodologies for pathogen and toxins: A Review. *Sensors* 2017, 17, 1–20.

Alghamdi, A. H. Applications of stripping voltammetric techniques in food analysis. *Arab. J. Chem.* 2010, 3, 1–7.

Aragáo-Catunda Jr., F. E.; de Araujo, M. F.; Granero, A. M.; Arévalo, F. J.; de Carvalho, M. G.; Zon, M. A.; Fernández, H. The redox thermodynamics and kinetics of flavonoid rutin adsorbed at glassy carbon electrodes by stripping square wave voltammetry. *Electrochim. Acta* 2011, 56, 9707–9713.

Arévalo, F. J.; González-Techera, A.; Zón, M. A.; González-Sapienza, G.; Fernández, H. Ultra-sensitive electrochemical immunosensor using analyte peptidomimetics selected from phage display peptide libraries. *Biosens. Bioelectron.* 2012, 32, 231–237.

Arévalo, F. J.; Granero, A. M.; Fernández, H.; Raba, J.; Zon. M. A. Citrinin (CIT) determination in rice samples using a micro fluidic electrochemical immunosensor. *Talanta* 2011, 83, 966–973.

Bard, A.L.; Faulkner, L. R. *Electrochemical Methods. Fundamentals and Applications*, 2nd ed. John Wiley & Sons: New York, 2001.

Borman, S. A. New electroanalytical pulse techniques. *Anal. Chem.* 1982, 54, 698-705.

Ceballos, C.; Fernández, H. Synthetic antioxidants determination in lard and vegetable oils by the use of voltammetric methods on disk ultramicroelectrodes. *Food Res. Int.* 2000a, 33, 357–365.

Ceballos, C.; Fernández, H. Synthetic antioxidants in edible oils by square wave voltammetry on ultramicroelectrodes. *J. Am. Oil Chem. Soc.* 2000b, 77, 731–735.

Ceballos, C. D.; Zon M. A.; Fernández, H. Using square wave voltammetry on ultramicroelectrodes to determine synthetic antioxidants in vegetable oils. *J. Chem. Educ.* 2006, 83, 1349–1352.

Chanique, G. D.; Arévalo, A. H.; Zon, M. A.; Fernández, H. Electrochemical reduction of patulin and 5-hydroxymethylfurfural in both neutral and acid non-aqueous media. Their electroanalytical determination in apple juices. *Talanta* 2013, 111, 85–92.

Daub, M. E. Cercosporin, a photosensitizing toxin from *Cercospora* species. *Phytopathology* 1982, 72, 370–374.

Díaz Nieto, C. H.; Granero, A. M.; García, D.; Nesci, A.; Barros, G.; Zon, M. A.; Fernández, H. Early warning system for detection of aflatoxin B_1 throughsterigmatocystin using a third generation biosensor. *Talanta* 2019, 194, 253–258.

Díaz Nieto, C. H.; Granero, A. M.; Zon, M. A.; Fernández, H. Novel electrochemical properties of an emergent mycotoxin: Sterigmatocystin. *J. Electroanal. Chem.* 2016, 765 (2016) 155–167.

Díaz Nieto, C. H.; Granero, A. M.; Zon, M. A.; Fernández, H. Sterigmatocystin: A mycotoxin to be seriously considered. *Food Chem. Toxicol.* 2018, 118, 460–470.

Díaz Toro, P. C.; Arévalo, F. J.; Fumero, M. V.; Zon, M. A.; Fernández, H. Very sensitive electrochemical sensor for moliniformin detection in maize samples. *Sens. Actuators B: Chem.* 2016, 225, 384–390.

Díaz Toro, P. C.; Arévalo, F. J.; Zon, M. A.; Fernández, H. Studies of the electrochemical behavior of moniliformin mycotoxin and its sensitive determination at pretreated glassy carbon electrodes in a non-aqueous medium. *J. Electroanal. Chem.* 2015, 738, 40–48.

Fernández, H. Mycotoxins quantification in the food system: Is there any contribution from electrochemical biosensors? Editorial. *J. Biosens. Bioelectron.* 2013, 4(3), 1–2.

Fernández, H.; Arévalo, F. J.; Granero, A. M.; Robledo, S. N.; Díaz Nieto, C. H.; Riberi, W. I.; Zon, M. A. Electrochemical biosensors for the determination of toxic substances related to food safety developed in South America: Mycotoxins and herbicides. *Chemosensors* 2017, 5, 23, 1–19.

Fernández, H.; Molina, P. G.; Arévalo, F. J.; Zon, M. A. *Micotoxinas: su rol en el sistema agroalimentario. Determinaciones electroanalíticas y por inmunoelectroanálisis.* In *Residuos urbanos e industriales ¿fuente de problemas o de oportunidades?* Vázquez, M. V.; Montoya Restrepo, J. (Eds.). RIARTAS Tecnológico de Antioquia: Medellín, Colombia, 2012a, p. 175.

Fernández, H.; Zon, M. A. *Recent Development and Applications of Electroanalytical Chemistry*. Research Signpost: Trivandrum, India, 2002.

Fernández, H.; Zon, M. A.; Molina, P. G.; Moressi, M. B.; Vettorazzi, N. R.; Arévalo, A. H.; Arévalo, F. J.; Granero, A. M.; Ramírez, E. A.; Perrotta, P.; Zachetti, V. G. L.; Chanique, G. *Electroanalytical properties of mycotoxins and their determinations in the agroalimentary system*. In *Mycotoxins: Properties, Applications and Hazards*; Melborn, B. J.; Greene, J. C. (Eds.). Nova Science Publishers, Inc.: New York, 2012b, p. 85.

Fleischmann, M.; Pons, S.; Rolison, D. R.; Schmidt, P. P. *Ultramicroelectrodes*. Datatech System, Inc., Science Publishers: USA, 1987.

González-Techera, A.; Zon, M. A.; Molina, P. G.; Fernández, H.; González-Sapienza, G.; Arévalo, F. J. Development of a highly sensitive noncompetitive electrochemical immunosensor for the detection of atrazine by phage anti-immunocomplex assay. *Biosens. Bioelectron.* 2015, 64, 650–656.

Granero, A. M.; Arévalo, F. J.; Fernández, H.; Zon, M. A. *Development of voltammetric techniques and sensors for the determination of resveratrol.* In *Resveratrol: Sources, Production and Health Benefits*. Delmas, D. (Ed.). Nova Science Publishers, Inc.: New York, 2013, p. 43.

Granero, A. M.; Fernández, H.; Agostini, E.; Zon, M. A. An amperometric biosensor for trans resveratrol determination in aqueous solutions by means of carbon paste electrodes modified with peroxidase basic isoenzymes from *Brassica napus*. *Electroanalysis* 2008, 20, 858–864.

Granero, A. M.; Fernández, H.; Agostini, E.; Zon, M. A. An amperometric biosensor based on peroxidases from *Brassica napus* for the determination of the total polyphenolic content in wine and tea samples. *Talanta* 2010, 83, 249–255.

Grumezescu, A. *Nanobiosensors*, 1st ed. Academic Press, London: United Kingdom, vol. 8, 2017.

International Agency for Research on Cancer (IARC), World Health Organization. Lyon, France, 2006. http://www.iarc.fr.

López, J. C.; Granero, A. M.; Robledo, S. N.; Zensich, M.; Morales, G. M.; Fernández, H.; Zon, M. A. Books of abstracts. In *Proceedings of the XIX Congreso Argentino de Fisicoquímica y Química Inorgánica*, Buenos Aires, Argentina, April 12–15, 2015. Asociación Argentina de Investigación Fisicoquímica: Buenos Aires, Argentina, 2015, P343.

López, J. C.; Zon, M. A.; Fernández, H.; Granero, A. M.; Robledo, S. N. Determination of thermodynamic and kinetic parameters of the enzymatic reaction between soybean peroxidase and natural antioxidants using chemometric tools. *Food Chem.* 2019a, 275, 161–168.

López, J. C.; Zon, M. A.; Fernández, H.; Granero, A. M. Development of an enzymatic biosensor for the determination of eugenol in dental materials. *Talanta* 2020, 210, 120647.

Marchiando, N. C.; Zon, M. A.; Fernández, H. Characterization of the surface redox process of adsorbed cercosporin (CER) at glassy carbon electrodes by anodic stripping square wave voltammetry. *Electroanalysis* 2003, 15, 40–48.

Marchiando, N. C.; Zon, M. A.; Fernández, H. Detection and quantification of cercosporin (CER) phytotoxin isolates from infected peanut leaves by using adsorptive stripping square-wave voltammetry. *Anal. Chim. Acta* 2005, 550, 199–203.

Maza, E.; Fernández, H.; Moressi, M.; Zon M. A. Electrochemical oxidation of fisetin flavonoid. Studies related to its adsorption at glassy carbon electrodes. *J. Electroanal. Chem.* 2012, 675, 11–17.

Maza, E. M.; Fernández, H.; Zon, M. A.; Moressi, M. B. Electrochemical determination of fisetin using gold electrodes modified with thiol self-assembled monolayers. *J. Electroanal. Chem.* 2017, 790, 1–10.

Mishra, G. K.; Barfidokht, A.; Tehrani, F.; Mishra, R. K. Food safety analysis using electrochemical biosensors. *Foods* 2018, 7, 1–11.

Molina, P. G.; Zon, M. A.; Fernández, H. Determination of electrochemical properties of the adsorbed zearalenone (ZEA) mycotoxin by using cyclic and square wave voltammetry. *Indian J. Chem.* 2003, 42A, 789–796.

Molina, P. G.; Zon, M. A.; Fernández, H. Novel studies about the electrooxidation of a deoxinivalenol (DON) mycotoxin reduction product adsorbed on glassy carbon and carbon paste electrodes. *Electroanalysis* 2008, 20, 1633–1638.

Molina, P. G; Zon, M. A.; Fernández, H. The redox kinetics of adsorbed ATX-I at carbon electrodes by anodic stripping square wave voltammetry. *Electroanalysis* 2000, 12, 791–798.

Molina, P. G.; Zon, M. A.; Fernández, H. The electrochemical behavior of theAltenuene mycotoxin and its acidic properties. *J. Electroanal. Chem.* 2002, 520, 94–100.

Moressi, M. B.; Andreu, R.; Calvente, J. J.; Fernández, H.; Zon, M. A. Improvement of alternariol monomethyl ether detection at gold electrodes modified with a dodecanethiol self-assembled monolayers. *J. Electroanal. Chem.* 2004, 570, 209–217.

Moressi, M. B.; Calvente, J. J.; Andreu Fondacabe, R.; Fernández; Zon, M. A. Electrooxidation of altertoxin (ATX-I) at gold electrodes modified by dodecanethiol self-assembled monolayers. Its quantitative determination. *J. Electroanal. Chem.* 2007, 605, 118–124.

Moressi, M. B.; Zon, M. A.; Fernández, H.; Rivas, G.; Solís, V. Amperometric quantification of alternaria mycotoxins with a mushroom tyrosinase modified carbon paste electrode. *Electrochem. Commun.* 1999, 1/10, 472–476.

Mustafa, F.; Andreescu, S. Chemical and biological sensors for food-quality monitoring and smart packaging. *Foods* 2018, 7, 168, 1–20.

Mustafa, F.; Hassan, R. Y. A.; Andreescu, S. Multifunctional nanotechnology-enabled sensors for rapid capture and detection of pathogens. *Sensors* 2017, 17, 1–28.

Osteryoung, J. Pulse voltammetry. *J. Chem. Educ.* 1983, 60, 296–298.

Perrotta, P. R.; Vettorazzi, N. R.; Arévalo, F. J.; Granero, A. M.; Chulze, S. N.; Zon, A. M.; Fernández, H. Electrochemical studies of ochratoxin A mycotoxin at gold electrodes modified with cysteamine self-assembled monolayers. Its ultrasensitive quantification in red wine samples. *Electroanalysis* 2011, 23, 1585–1592.

Perrotta, P. R.; Vettorazzi, N. R.; Arévalo, F. J.; Zon, M. A.; Fernández, H.Development of a very sensitive electrochemical magneto immunosensor for direct determination of ochratoxin A in red wine. *Sens. Actuators B Chem.* 2012, 162, 327–333.

Pierini, G. D.; Granero, A. M.; Di Nezio, M. S.; Centurión, M. E.; Zon, M. A.; Fernández, H. Development of an electroanalytical method for the determination of lead in Argentina raw propolis based on bismuth electrodes. *Microchem. J.* 2013, 106, 102–106.

Pierini, G. D.; Robledo, S. N.; López, J. C.; Tesio, A. Y.; Fernández, H.; Granero, A. M.; Zon, M. A. *Glassy carbon electrodes: Studies regarding antioxidant reaction mechanisms and their electroanalytical applications.* In *Advances in Materials Science Research*; Wythers M. (Ed.). Nova Science Publishers, Inc.: New York, 2017, vol. 29, p. 189.

Pierini, G. D.; Robledo, S. N.; Zon, M. A.; Di Nezio, M. S.; Granero, A. M.; Fernández, H. Development of an electroanalytical method to control quality in fish samples based on edge plane pyrolytic graphite electrode. Simultaneous determination of hypoxanthine, xanthine and uric acid. *Microchem. J.* 2018, 138, 58–64.

Pividori, M. I.; Alegret, S. Electrochemical biosensors for food safety. *Contrib. Sci.* 2010, 6, 174–191.

Pokorny, J.; Yanishlieva, N.; Gordon, M. *Antioxidantes de los Alimentos. Aplicaciones Prácticas*; Acribia, S. A. (Ed.). Zaragoza: España, 2001.

Queiros, M. A.; Daschbach, J. L.; Montenegro, M. I. *Microelectrodes: Theory and Applications*; NATO ASI Series. Academic Publisher: Dordrecht, 1990, vol. 197.

Ramírez, E. A.;Granero, A. M.;Zon, M. A.;Fernández, H. Development of an amperometric biosensor based on peroxidases from *Brassica napus* for the determination of ochratoxin A content in peanut samples. *J. Biosens. Bioelectron.* 2011, S3, 1–6.

Ramírez, E. A.; Molina, P. G.; Zon, M. A.; Fernández, H. Development of an electroanalytical method for the quantification of zearalenone (ZEA) in maize samples. *Electroanalysis* 2005, 17, 1635–1640.

Ramírez, E. A.; Zon, M. A.; Jara Ulloa, P. A.; Squella, J.A.; Nuñez Vergara, L.; Fernández, H. Adsorption of ochratoxin A (OTA) anodic oxidation product on glassy carbon electrodes in highly acidic reaction media: Its thermodynamic and kinetics characterization. *Electrochim. Acta* 2010, 55, 771–778.

Rhazi, M. E.; Majid, S.; Elbasri, M.; Salih, F. E.; Oularbi, L.; Lafdi, K. Recent progress in nanocomposites based on conducting polymer: application as electrochemical sensors. *Int. Nano Lett.* 2018, 8, 79–99.

Riberi, W. I. Estudios electroquímicos de micotoxinas del género *Fusarium*. Inmovilización de biomoléculas sobre electrodos nanoestructurados y desarrollo de inmunosensores. Aplicaciones analíticas. PhD Dissertation, Universidad Nacional de Río Cuarto, Río Cuarto, Córdoba, Argentina, 2019.

Riberi, W. I.; Tarditto, L. V.; Zon, M. A.; Arévalo, F. J.; Fernández, H. Development of a very sensitive electrochemical immunosensor to determine zearalenone in maize samples. *Sens. Actuators B: Chem.* 2018, 254C, 1271–1277.

Riberi, W. I.; Zon, M. A.; Fernández, H.; Arévalo, F. J. Impedimetric immunosensor to determine patulin mycotoxin in fruit juice using a glassy carbon electrode modified with graphene oxide. *Microchem. J.*, 2019.

Robledo, S. N.; Granero, A. M.; Zon, M. A.; Fernández, H. *Electrochemical Determination of Antioxidants in Edible Vegetable Oils*. LAP LAMBERT Academic Publishing: Saarbrücken, Germany, 2014.

Robledo, S. N.; Pierini, G. D.; Díaz Nieto, C. H.; Fernández, H.; Zon, M. A. Development of an electrochemical method to determine phenolic monoterpenes in essential oils. *Talanta* 2019, 196, 362–369.
Robledo, S. N.; Tesio, A. Y.; Ceballos, C. D.; Zon, M. A.; Fernández, H. Electrochemical ultra-micro sensors for the determination of synthetic and natural antioxidants in edible vegetable oils. *Sens. Actuators B: Chem.* 2014, 192, 467–473.
Robledo, S. N.; Zachetti, V. G. L.; Zon, M. A.; Fernández, H. Quantitative determination of tocopherols in edible vegetable oils using electrochemical ultra-microsensors combined with chemometric tools. *Talanta* 2013, 116, 964–971.
Robledo, S. N.; Zon, M. A.; Ceballos, C. D.; Fernández, H. Qualitative and quantitative electroanalysis of synthetic phenolic antioxidant mixtures in edible oils based on their acid-base properties. *Food Chem.* 2011, 127, 1361–1369.
Segneanu, A.; Grozescu, I.; Cepan, C.; Velciov, S. Significance of food quality on human health. *Appl. Food Sci. J.* 2018, 2(2), 17–19.
Soriano del Castillo, J. M. *Micotoxinas en alimentos*. Díaz de Santo: Spain, 2007.
Tesio, A. Y.; Granero, A. M.; Vettorazzi, N. R.; Ferreyra, N. F.; Rivas, G. A.; Fernández, H.; Zon, M. A. Development of an electrochemical sensor for the determination of the flavonoid luteolin in peanut hull samples. *Microchem. J.* 2014, 115, 100–105.
Tesio, A. Y.; Granero, A. M.; Fernández, H.; Zon, M. A. Characterization of the surface redox process of adsorbed morin (MO) at glassy carbon electrodes. *Electrochim. Acta* 2011, 56, 2321–2327.
Tesio, A. Y.; Robledo, S. N.; Granero, A. M.; Fernández, H.; Zon, M. A. *Electroanalytical determinations of luteolin.* In *Luteolin: Natural Occurrences, Therapeutic Applications and Health Effects*; Dwight, A. J. (Ed.). Nova Science Publishers, Inc.: New York, 2015, p. 73.
Tesio, A. Y.; Robledo, S. N.; Fernández, H.; Zon M. A. Electrochemical oxidation of butein at glassy carbon electrodes. *Bioelectrochemistry* 2013, 91, 62–69.
Vettorazzi, N. R.; Zon, M. A.; Molina, P. G.; Granero, A. M.; Arévalo, F. J.; Robledo, S. N.; Díaz Toro, P. C.; Díaz Nieto, C. H.; Fernández, H. *Métodos de análisis de contaminantes*. Red Iberoamericana de Aprovechamiento de Aguas y Suelos Contaminados (RIARTAS)—Programa Iberoamericano—CYTED: Medellín, Colombia, 2014, p. 1.
Viswanathan, S.; Radecka, H.; Radecki, J. Electrochemical biosensors for food analysis. *Monatsh. Chem.* 2009, 140, 891–899.
Vitaglione, P.; Sforza, S.; Galaverna, G; Ghidini, C.; Caporaso, N.; Vescovi, P. P.; Fogliano, V.; Marchelli, R. Bioavailability of trans-resveratrol from red wine in humans. *Mol. Nutr. Food Res.* 2005, 49, 495–504.
Wang, J. *Analytical Electrochemistry*, 3rd ed. Wiley-VCH: USA, 2006.
Zachetti, V. G. L.; Granero, A. M.; Robledo, S. N.; Zon, M. A.; DaRocha Rosa, C. A.; Fernández, H. Electrochemical reduction of the mycotoxin citrinin at bare glassy carbon electrodes and modified with multi-walled carbon nanotubes in a non-aqueous reaction medium. *J. Braz. Chem. Soc.* 2012, 23, 1131–1139.
Zachetti, V. G. L.; Granero, A. M.; Robledo, S. N.; Zon, M. A.; Fernández, H. Development of an amperometric biosensor based on peroxidases to quantify citrinin in rice samples. *Bioelectrochem.* 2013, 91, 37–43.
Zeng, L.; Peng, L.; Wu, D.; Yang, B. *Electrochemical Sensors for Food Safety*. IntechOpen, London: United Kingdom, 2018. doi: http://dx.doi.org/10.5772/intechopen.82501

Zon, M. A.; Ceballos, C.; Molina, P. G.; Marchiando, N. C.; Moressi, M. B.; Fernández, H. *Application of electroanalytical techniques for the quantitative determination of synthetic antioxidants and micotoxins produced by fungus*. In *Recent Research Developments in Electroanalytical Chemistry*. Transworld Research Network: Trivandrum, India, 1999, vol. 1, p. 115.

Zon, M. A.; Marchiando, N. C.; Fernández, H. A study of the bielectronic electro-reduction of cercosporin phytotoxin in highly acidic non-aqueous medium. *J. Electroanal. Chem.* 1999, 465, 225–233.

Zon, M. A.; Vettorazzi, N. R.; Moressi, M. B.; Molina, P. G.; Granero, A. M.; Arévalo, F. J.; Robledo, S. N.; Fernández, H. *Voltammetric techniques applied on organic compounds. Applications to some compounds related to agroalimentary and health systems*. In *Voltammetry: Theory, Types and Applications*; Saito, Y.; Kikuchi, T. (Eds.). Nova Science Publishers, Inc.: New York, 2014, p. 85.

CHAPTER 6

POMEGRANATE SEEDS AS A POTENTIAL SOURCE OF PUNICIC ACID: EXTRACTION AND NUTRACEUTICAL BENEFITS

JUAN M. TIRADO-GALLEGOS[1], R. BAEZA-JIMÉNEZ[2], JUAN A. ASCACIO-VALDÉS[3], JUAN C. BUSTILLOS-RODRÍGUEZ[4], and JUAN BUENROSTRO-FIGUEROA[2*]

[1]*School of Animal Sciences and Ecology, Autonomous University of Chihuahua, 31453 Chihuahua, Chihuahua, México*

[2]*Research Center in Food and Development, A.C. 33089 Cd. Delicias, Chihuahua, México*

[3]*Bioprocesses & Bioproducts Group, Food Research Department, School of Chemistry, Autonomous University of Coahuila, 25280 Saltillo, Coahuila, México*

[4]*Research Center in Food and Development, A.C. 31570 Cd. Cuauhtémoc, Chihuahua, México*

[*]*Corresponding author. E-mail: jose.buenrostro@ciad.mx*

ABSTRACT

Pomegranate (*Punica granatum* L.) is a fruit cultivated under diverse climatic conditions, being an edible fruit with great adaptability and flexibility around the world. Its consumption has been related to health benefits. Pomegranate seeds are an interesting component of this fruit due to their oil content (up to 50 wt.%), which have been distinguished for their pharmaceutical applications and nutraceutical properties. The main fatty acid residue identified in the oil is punicic acid (PuA) (≈80%, with respect to the total fatty acid content).

PuA can exert important bioactivities such as anticancer, hypolipidemic, antidiabetes, antiobesity, antioxidant, anti-inflammatory, among others. One of the most potent sources of PuA is pomegranate seed, a byproduct obtained during the industrialization of this fruit. In this sense, one of the strengths in PuA research is the revalorization of this agroindustrial residue for the further recovery of such bioactive fatty acids. On the other hand, new and novel applications for PuA can be developed. Therefore, in the present chapter, it will be fully detailed the research fields on PuA, its application in pharmaceutical, cosmetic, and food industries, its different existing sources, its extraction methods, the analytical techniques for its identification, and its nutraceutical benefits in food and human health.

6.1 INTRODUCTION

A native of the Middle East, pomegranate (*Punica granatum* L.) fruit is categorized within the group of berries. There is a fruit commonly known as "romã," "romazeira," "mangrano," and "granado" in Latin American countries (dos Santos Souza et al., 2018). The main countries that are producers in the world are Afghanistan, Iran, India, the United States, and Spain. About the total area and total production of pomegranate, no clear data is available. However, total production has been estimated to be around three million tons (FAO, 2014).

A rapid increase in its cultivation and consumption has been observed due to the several health properties (Ascacio-Valdés et al., 2011). Then, pomegranate is not only consumed as food or as decoration in food but also used in traditional herbal medicine and recently incorporated in food, cosmetic, and pharmaceutical industries in the form of juices, nectars, teas, food supplements, wine, liquor, pills, creams, facial, and body oils (Mercado-Silva et al., 2011).

Derived from its processing, different residues can be obtained, such as peels and seeds. Due to their composition, these materials can be used to obtain compounds of industrial interest, such as phenolic compounds, enzymes, and oil. Pomegranate seeds are an excellent source to obtain oil, which is rich in unsaturated fatty acids, particularly punicic acid (PuA), an isomer of α-linolenic acid (Holic et al., 2018).

For the aforementioned, the recent advances in PuA research fields will be described as well as its sources, extraction and quantification methods, industrial applications, and nutraceutical benefits on human health.

6.2 CHEMICAL COMPOSITION

The pomegranate fruit and juice compositions are clearly affected by several factors such as variety, geographical location, agronomical practices, harvest time, processing, and storage conditions (Fischer et al., 2013). Pomegranate parts including arils, seeds, rind, flowers, bark, and roots, contain a wide range of bioactive compounds, including polyphenols (punicalagin, punicalin, ellagic acid, flavonoids, anthocyanins), fatty acids, sterols, terpenoids, and alkaloids (Ascacio-Valdés et al., 2011; Amri et al., 2017).

Pomegranate fruit has an edible fraction consisted of the arils, which corresponds up to 60% of total weight fruit. These arils contain 80% of juice and 20% of seeds (Erkanand-Kader, 2011). According to the United States Department of Agriculture (USDA, 2016), 100 g of whole fruit (Wonderful variety) and its juice contain, respectively (%), water 77.9 and 86, protein 1.67 and 0.15, total lipid 1.17 and 0.29, carbohydrate 18.7 and 13.1, total dietary fiber 4 and 0.1, and total sugars 13.7 and 12.7. Also, it contain minerals such as Ca, Fe, Mg, P, K, Na, and Zn, as well as vitamins C, B_1, B_2, B_3, B_6, B_9, E, and K. Several studies have highlighted the elevated phenolic content of pomegranate juice, compared to those reported in red wine and other berries (cranberry, strawberry, blueberry), exhibiting an antioxidant activity three-fold higher than those of red wine and green tea (Gil et al., 2000).

This antioxidant activity is due to the presence of several phenolic compounds, such as punicalagin, punicalin, anthocyanins, and ellagic acid (Ascacio-Valdés et al., 2011). However, changes in phenolic profile occur during juice extraction: by manual pressing, where arils are separated from peel, albedo, and carpel membranes and then pressed through a mesh; and by a squeezer, where a press or squeezer is used to compress the whole fruit, which allows other phenolic compounds present in the peel and albedo to migrate to juice, resulting in a juice with an enriched phenolic compounds content (Calani et al., 2013; Nuncio-Jáuregui et al., 2015).

6.3 POMEGRANATE CONSTITUENTS

Pomegranate fruit is composed of three main constituents: arils, peel, and seeds, each one with important molecules present, responsible for several biological properties.

6.3.1 ARILS

The edible fraction is the aril, corresponding to 55%–60% of total pomegranate weight and containing about 80% juice, and the rest is the seed (Erkan and Kader, 2011). This fraction is consumed directly, or once the seeds have been separated, the juice is used to prepare fresh or alcoholic beverages, jams, and jellies and to obtain flavors and colorants (Szychowski et al., 2015). Arils are used as an ingredient in cuisine in baked goods, energy bars, yogurt, ice cream, and salad dressings and their dehydrated form for easy conservation (Horuz and Maskan, 2015). Pomegranate arils are composed of fructose, glucose, pectin, citric acid, malic acid, and bioactive compounds, namely, phenolics and flavonoids.

The consumption of juice and its products has increased during last year due to the phenolic compounds present, which exhibit several bioactivities that have attractive pharmaceutical applications: anticancer and antimicrobial activity, antiatherogenic effect, control of cardiovascular diseases, diabetes, hypercholesterolemia, hypertension, and obesity (Al-Muammar and Khan, 2012; Bassiri-Jahromi, 2018; Çam and Hışıl, 2010; Seeram et al., 2005).

The juice is well known as an excellent source of antioxidants (polyphenols, tannins, and anthocyanins, including vitamins C and E, coenzyme Q10, and lipoic acid). However, although arils have a high content of anthocyanins, it has been reported that the nonedible fractions (peels and seeds) have an elevated content of phenolic compounds, mainly ellagitannins, which have been associated with several health beneficial properties (Fischer et al., 2011, 2013; Wasila et al., 201).

6.3.2 PEEL

Pomegranate peel has a high antioxidant activity that is related to the high phenolic content, mainly punicalagin, a hydrolyzable tannin unique to the pomegranate fruit (Ascacio-Valdés et al., 2011). Hydrolyzable tannins represent around 80%–85% of the total phenol content in the peel, and they can act as antitumoral or hepatotoxic agents, improving cardiovascular health; they also exert antiviral, antidiabetic, and anti-inflammatory activities and contribute to oral and skin health (Viuda-Martos et al., 2010). Studies on the revalorization of pomegranate peel have been developed to produce enzymes, polysaccharides, and bioactive compounds using fermentative processes (Akhtar et al., 2015; Çam and İçyer, 2015; Sepúlveda et al., 2012; Zhu and Liu, 2013).

6.3.3 SEEDS

Pomegranate seeds comprise 3% of fruit in weight, with lower polyphenol content. However, besides its important physiological and nutritional role, the seed is a potential source of oil. The oil content varies from 7% to 20% in seed weight and it depends on the cultivar (Singh et al., 2002). Pomegranate seed oil (PSO) has a high content (65%–80%) of conjugated fatty acids such as linoleic, linolenic, and other lipids such as PuA, oleic, stearic, and palmitic acids (Özgül-Yücel, 2005). Conjugated fatty acids play an important preventive role in cardiovascular disease, as well as in the treatment of other health problems since they promote the reduction of both total and high-density lipoprotein cholesterol (Al-Muammar and Khan, 2012). However, the most important fatty acid is 9-*trans*,11-*cis*,13-*trans*-octadecatricnoic acid, known as PuA, with content of ~80% (with respect to the total amount of fatty acids), depending on the extraction conditions (Abbasi et al., 2008). Pomegranate seeds are rich in phytosterols and other components, such as proteins, fibre, vitamins, and minerals; polyphenols and isoflavones also contribute to the overall spectrum of health benefits.

6.4 POMEGRANATE SEED OIL

The lipid fraction of pomegranate seeds is of particular interest due to the polyunsaturated essential fatty acid content (Sassano et al., 2009). PSO is rich in fatty acids (80%) (Holic et al., 2018) of 18-carbon chains with three alternating double bonds. These molecules, also known as trienoic acids, present higher physiological activity than dienoic fatty acids (Kýralan et al., 2009; Aruna et al., 2016). The oil content in the seeds can vary from 12% to 20% of the total seed weight (Barizão et al., 2015; Bedel et al., 2017; Holic et al., 2008; Meerts et al., 2009; Karimi et al., 2017). Kýralan et al. (2009) evaluated the seed oil content in 15 pomegranate cultivars of Turkey. The PSO varied between 13.95% and 24.13% and the PuA content ranges from 70.42% to 76.17% on dry basis (db). In another study, Soetjipto et al. (2010) examined the fatty acid composition of red and purple pomegranate varieties from Indonesia. The total oil content was higher in red pomegranate (12.8%) than purple pomegranate (10.3%, db). The PuA oil contents of the total lipids of purple and red pomegranate were 0%–25% and 9%–16% (db), respectively. Neutral lipids of red pomegranate showed higher PuA contents (54%–75%) than purple pomegranates (14%–55%). After PuA, the second most common fatty acid is linoleic acid (0.7%–24.4%), followed by oleic acid (0.4%–17.7%), stearic acid (2.8%–16.7%), and palmitic acid (0.3%–9.9%). The oil content is strongly influenced by the cultivar, geographical location, growing conditions, and maturity stage (Holic et al., 2018; Meerts et al., 2009; Miranda et al., 2013). The fatty acid profile of PSO reported by researcher groups is shown in Table 6.1. There is a great variation between the concentrations reported for the fatty acid composition. These differences have been attributed to the variety of fruits and maturity extraction methods (Topkafa et al., 2015). In general, PSO is a rich source of polyunsaturated fatty acids, mainly PuA, which constitute up to 80% of total fatty acids (Schneider et al., 2012; Shaban et al., 2013).

PSO also contains phytosterols such as beta-sitosterol, campesterol, and stigmasterol, which contribute to health benefits. Đurđević et al. (2018), examined the concentration of tocopherols (α, γ, δ) and β-tocotrienol and carotenoids in PSO. The authors reported 128.8 mg/100 g of γ-tocopherol, 4.3 mg/100 g of α-tocopherol, 2.53 mg/100 g of δ-tocopherol, and 2.71 mg/100 g of β-tocotrienol. The γ-tocopherol is characterized as the tocol most resistant to oxidation, α-tocopherol is a good antioxidant. Carotenoids were detected in trace amounts; the most abundant was (E)-β-carotin (0.15 mg/100 g). Also, it has been reported that PSO contains phytosterols (beta-sitosterol, campesterol, and stigmasterol) with concentrations of 4089–6205 mg/kg of PSO (Kaufman and Wiesman, 2007). Other minor components of PSO are cerebroside, isoflavone genistein, the phytoestrogen coumestrol, the sex steroid estrone, and nonsteroidal estrogens like daidzein (a glucoside), genistein (a glycone), and coumestrol (Karimi et al., 2017). Many of these components contribute to the overall spectrum of the biological activity of PSO. Regarding the physicochemical quality of pomegranate oil, Costa et al. (2019) evaluated the quality of commercial cold-pressed PSO from Turkey and Israel. The authors referred to a peroxide index between 0.91 and 2.69, which is lower than the maximum acceptable value (15). The acid value was between 1.8 and 4.38 expressed as % PuA, which was below and slightly above the maximum acceptable value (4). On the other hand, the oil stability

TABLE 6.1 Fatty Acid Composition (%) of PSO

Fatty acid	Siano et al. (2015)	Topkafa et al. (2015)	Białek et al. (2018)	Costa et al. (2019)
Myristic acid (14:0)	nr	0.01 ± 0.01	nr	nr
Palmitic (C16:0)	4.87 ± 1.32	2.49 ± 0.01	4.5 ± 0.20	1.99 ± 0.07
Palmitoleic (16:1)	nr	0.01 ± 0.00	nr	nr
Heptadecanoic (C17:0)	0.14 ± 0.02	0.03 ± 0.00	0.5 ± 0.10	nr
Heptadecenoic acid (C17:1)	nr	0.06 ± 0.00	nr	nr
Stearic (C18:0)	2.78 ± 0.95	1.82 ± 0.01	2.9 ± 0.20	1.64 ± 0.08
Oleic (C18:1)	8.15 ± 1.83	3.91 ± 0.02	12.2 ± 0.60	4.00 ± 0.16
Linoleic (C18:2)	9.59 ± 2.03	4.73 ± 0.00	9.7 ± 0.40	4.67 ± 0.02
Linolenic (18:3 n3)	0.40 ± 0.05	nr	1.1 ± 0.00	nr
Arachidic (C20:0)	0.13 ± 0.05	0.38 ± 0.00	0.5 ± 0.00	0.54 ± 0.05
Eicosenoic (C20:1)	nr	0.53 ± 0.03	1.2 ± 0.70	0.69 ± 0.09
Eicosadienoic acid (C20:2)	nr	0.13 ± 0.04	nr	nr
Eicosapentaenoic (20:5)	0.05 ± 0.01	nr	nr	nr
Behenic acid (C22:0)	nr	0.05 ± 0.04	0.2 ± 0.70	nr
Tricosylic acid (C23:0)	nr	0.20 ± 0.00	4.4 ± 4.00	nr
Punicic acid (C18:3)	55.27 ± 2.58	76.57 ± 0.02	15.1 ± 0.90	60.62 ± 5.40
Catalpic acid	1.61 ± 0.35	6.47 ± 0.03	nr	4.79 ± 0.36
β-Oleostearic acid	nr	1.41 ± 0.03	nr	1.41 ± 0.17
CLnA1	nr	nr	13.1 ± 1.4	0.64 ± 0.06
CLnA2	nr	nr	3.9 ± 1.2	0.32 ± 0.07
CLnA3	nr	nr	13.6 ± 3.1	0.41 ± 0.04
CLnA4	nr	nr	4.7 ± 0.80	0.32 ± 0.03
Lignoceric (C24:0)	1.84 ± 0.87	nr	nr	nr

CLnA1–CLnA4: unidentified isomers of conjugated linolenic acids, nr: not reported.

index (0.10–0.22 h) was influenced by the concentration of conjugated isomers of linoleic acid regardless of the content of antioxidants in oils. In another study, Khoddami et al. (2014) determined some physicochemical properties of PSO from pomegranate extracted by cold pressing from the variety Torshe Malas Iran. The thermal behavior of the three oils was determined using differential scanning calorimetry, which registered a melting point of −12.70 °C. The PSO had a low peroxide value (4.67 meq/kg), low free fatty acid content (0.65% as PuA), and high total phenolic content (10.44 mg GAE/g sample). Based on their results, the authors concluded that the extracted crude oil showed good quality. In general, PSO obtained by cold pressing showed acceptable quality indexes, which depend on the type of pomegranate.

6.5 PUNICIC ACID (PUA)

6.5.1 CHEMICAL STRUCTURE

Conjugated linolenic acid (CLnA) isomers have attracted the attention of researchers for their benefits in human health (Miranda et al., 2013; Vroegrijk et al., 2011). The conjugated fatty acids are positional and geometric isomers of polyunsaturated fatty acids with conjugated double bonds (Yuan et al., 2009). One of the CLnA isomers with potential applications is PuA, an omega-5 fatty acid. PuA is an isomer of α-linolenic acid, which is characterized by three alternating double bonds (–CH=CH–CH=C–), these are called triene-type (Figure 6.1). According to the International Union of Pure and Applied Chemistry, the chemical name of this essential fatty acid is *9Z,11E,13Z-octadeca-9,11,13-trienoic acid*, with three double bonds (*cis9, trans11, and cis13*) (Miranda et al., 2013; Shabbir et al., 2017; Vroegrijk et al., 2011).

6.5.2 NATURAL BIOSYNTHESIS AND METABOLISM

Biosynthesis of PuA was studied by Hornung et al. (2002), who proposed (11,14)-linoleoyl desaturase activity for the conversion of a cis-double-bond at position δ12 into a *cis–trans* double-bond system. Two cDNAs from pomegranate seeds encoding for these enzymes were cloned and expressed in *Saccharomyces cerevisiae*. Fatty acids (0.02%, w/v) were added to the recombinant yeast culture and the analysis of the fatty acids produced by the yeast metabolism evidenced that one of the cDNA codes for δ12-acyl-lipid-desaturase, while the other codes for (1,4)-acyl-lipid-desaturase that converts the cis-double-bond at the δ12-position of linoleic acid or γ-LnA, but not α-LnA, into a conjugated *cis–trans* double-bond acids. In a later study, Iwabuchi et al. (2003) reported two cDNAs, one isolated from *Trichosantheskirilowii* (TkFac) and other isolated from *Punica granatum* (PgFac), that encode a class of conjugases associated with the formation of *trans*-11, *cis*-13 double bonds. The expression of both cDNAs in *Arabidopsis* seeds promoted accumulation of PuA up to ~10% (wt) of the total seed oil. Moreover, the expression of TkFac and PgFac in yeast grown without the addition of exogenous fatty acids, TkFac and PgFac expression resulted in PuA accumulation accompanied by 16:2 δ9cis,12cis and 18:2 δ9cis,12cis production. Therefore, TkFac and PgFac are bifunctional enzymes having both conjugase and δ12-oleate desaturase activities. Also, 16:2 δ9cis,12cis and 18:3 δ9cis,12cis,15cis, as well as 18:2 δ9cis,12cis, may be potential substrates for the conjugases to form trans-δ11 and cis-δ13 double bonds.

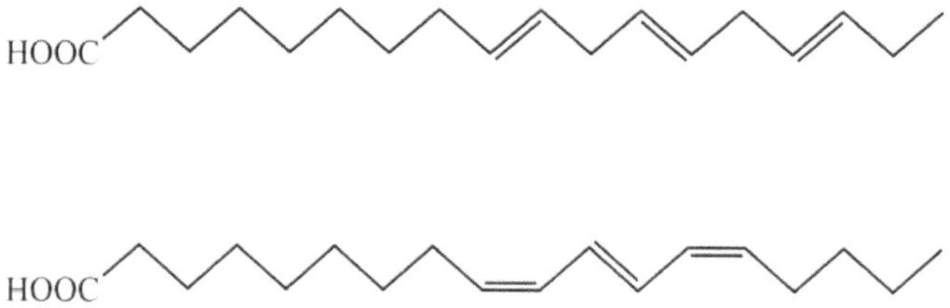

α-linolenic acid

Punicic acid

FIGURE 6.1 Linoleic acid and PuA.

On the other hand, with respect to studies conducted to elucidate the metabolism and bioavailability of PuA in living organisms, most of them have employed rats. Tsuzuki et al. (2006) examined the absorption and metabolism of α-eleostearic acid (α-ESA) of PSO in rat intestine using a lipid absorption assay in lymph from the thoracic duct. PuA is slowly absorbed in the intestine without changes and another part is rapidly converted to CLnA in the intestine of rats (Figure 6.2). The absorbed PuA can be converted into linoleic acid using a saturation reaction in various rat tissues, such as the liver, adipose tissue, plasma, and brain (Tsuzuki et al., 2006; Yuan et al., 2009). In another *in vivo* study, Yuan et al. (2009) examined the incorporation and metabolism of PuA in healthy young humans. The authors reported that PuA could be incorporated into the plasma and red blood cell membranes, and this was associated with the increasing proportion of CLnA. One of the enzymes that may be involved in the process of saturation of double bond in the carbon 13 is the nicotinamide adenine dinucleotide phosphate (NADP) since it is the only enzyme known to date that can rearrange fatty acids with conjugated double bonds, specifically trienoic acids (Schneider et al., 2012).

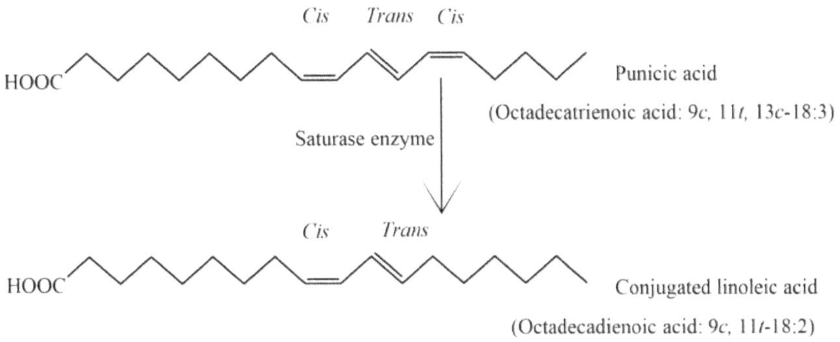

FIGURE 6.2 Metabolism of PuA.

6.6 EXTRACTION METHODS OF PSO

The extraction procedures of PSO include the use of a solvent in Soxhlet, stirring, microwave or ultrasonic irradiation, supercritical CO_2 extraction, and superheated solvent extraction. Other techniques provide an ecofriendly process, such as cold and hot pressing, where no solvent is used (Table 6.2). Oil yield depends on the efficiencies of the different extraction methods and their conditions, achieving yields from 7% to 59.1%, with 53.60%–82.90% PuA content. With respect to solvents, hexane has been mainly utilized and in minor grade chloroform, petroleum benzene, petroleum ether, methanol, and ethanol. Several factors affect the yield of PSO, such as temperature, solvent/mass ratio, particle size, extraction time, among others. The highest PSO yields have been reached with hexane, followed by chloroform, petroleum ether, Folch, and ethanol (Aruna et al., 2018).

The use of a solvent is necessary to suspend the seed powder and also for an improved mass transfer during agitation. The use of greater solvent/sample ratios (s/s) contributes to obtaining high yields of PSO, which occurs with Soxhlet extraction. This technique uses s/s ratios ranging from

TABLE 6.2 Effect of Extraction Method on Oil Yield from Pomegranate Seed and its PuA Content

Method	Conditions (solvent, temperature, particle size, time, solvent/mass ratio, pressure, flow rate)	Yield Oil (% w/w)	PuA (%)	Reference
Soxhlet	Hexane, 6 h, 31 mL/g	18.70	81.40	Abbasi et al. (2008)
	Petroleum benzene, 6 h, 31 mL/g	18.60	81.50	Abbasi et al. (2008)
	Hexane, 68 °C, 0.50–0.76 mm, 20 h	22.31	–	Ahangari and Sargolzaei (2012)
	Hexane, 110 °C, 8 h, 22 mL/g	34.70	–	Çavdar et al. (2017)
	Hexane, 68 °C, 6 h, 25 mL/g	26.8	77.23	Ghorbanzadeh and Rezaei (2017)
	Hexane, 80 °C, 0.25 mm, 24 h (8 h every day), 20 mL/g	17.94	81.69	Eikani et al. (2012)
	Hexane, 60 °C, 3 h, 30 mL/g	20.50	–	Tian et al. (2013)
Stirring	Hexane, 4 h, 4 mL/g	13.00	81.50	Abbasi et al. (2008)
	Petroleum benzene, 46 h, 4 mL/g	13.00	82.1	Abbasi et al. (2008)
	Water, 63 °C, pH 5.0, 2.2 mL/g, 6.25 h, 100 rpm	19.30	81.40	Ghorbanzadeh and Rezaei (2017)
Cold-pressed	Hexane, 25 °C, 8 h, 10 mL/g	17.50	–	Çavdar et al. (2017)
	Hexane, 80 °C, 0.25 mm, 72 h, 20 g seeds, 10 ton	4.29	69.79	Eikani et al. (2012)
	500 g, 26 °C, 6.25 kg/h	7.00	77.03	Ghorbanzadeh and Rezaei (2017)
Hot pressed	500 g, 50 °C, 6.25 kg/h	8.60	77.17	Ghorbanzadeh and Rezaei (2017)
Super-heated hexane	Hexane, 80 °C, 0.25 mm, 2 h, 20 bar, 1 mL/min	22.18	70.73	Eikani et al. (2012)
Supercritical fluid	Propane, 12 Mpa, 60 °C, 80 min	17.02	–	Ahangari and Sargolzaei (2012)
	CO_2, 40 °C, 30 MPa	13.06	–	Ahangari and Sargolzaei (2012)
	CO_2, 50 °C, 4 mL/g, 30 MPa, 2 h	–	61.58	Liu et al. (2007)
	CO_2, 50 °C, 4 mL/g, 30 MPa, 30 min	–	65.64	Liu et al. (2007)
	CO_2, 47 °C, 2 h, 250 g of seed, 38 MPa, 21 L/h	15.72	–	Tian et al. (2013)

TABLE 6.2 *(Continued)*

Method	Conditions (solvent, temperature, particle size, time, solvent/mass ratio, pressure, flow rate)	Yield Oil (% w/w)	PuA (%)	Reference
Supercritical fluid	CO_2, 47 °C, 37.9 MPa, 0.3 kg/h	–	53.60	Đurđević et al. (2018)
Microwave	Hexane, 10 min, 200 W, 4 mL/g	14.70	81.70	Abbasi et al. (2008)
	Hexane, 10 min, 800 W, 4 mL/g	15.60	82.90	Abbasi et al. (2008)
	Petroleum benzene, 10 min, 200 W, 4 mL/g	15.00	81.10	Abbasi et al. (2008)
	Petroleum benzene, 10 min, 800 W, 4 mL/g	15.80	82.20	Abbasi et al. (2008)
	Hexane, 0.125–0.450 mm, 5 min, 176 W, 10 mL/g	34.91	–	Çavdar et al. (2017)
Ultrasound	Hexane, 45 min, 4 mL/g	16.00	82.00	Abbasi et al. (2008)
	Petroleum benzene, 45 min, 4 mL/g	15.70	81.70	Abbasi et al. (2008)
	Hexane, 20 °C, 0.2 mm, 20 mL/g, 60 % amplitude	59.10	–	Goula (2013)
	Petroleum ether, 40 °C, 36 min, 140 W, 10 mL/g	25.11	65.00	Tian et al. (2013)

20 to 31 mL/g, compared to the other extraction methods that use s/s ratios of 4 mL/g. However, when an s/s ratio of 20 mL/g is used for ultrasound-assisted extraction, higher PSO yields are obtained as comparison to Soxhlet's (Table 6.2).

Particle size also affects PSO yield because mass transfer distance decreases and surface area increases for a minor particle size. Therefore, reduction in particle size increases the amount of sample exposed both to extraction by solvent and to the ultrasonically induced cavitation or microwave radiation (Goula, 2013).

Although the solubility, viscosity, and diffusivities are improved by the increased temperature of extraction, higher temperatures lead to protein coagulation, affecting negatively the PSO yield (Ghorbanzadeh and Rezaei, 2017). Conversely, for ultrasound-assisted oil extraction, at higher temperatures, vapor pressure increases, which exerts a great influence on the occurrence and intensity of acoustic cavitation (Goula, 2013). The use of microwave requires a lower amount of solvent and energy requirement, and provides a more effective extraction, because of the rapid heating and destruction of biological cell structure, releasing the oil (Çavdar et al. 2017).

Ghorbanzadeh and Rezaei (2017) reported the optimization of an aqueous extraction process for PSO. The authors mentioned that pH improves the solubility of the surrounding proteins, based on their isoelectric points. The yield obtained was 7% lower than the obtained with hexane extraction but higher than those obtained with cold and hot press. Also, PuA was obtained in higher concentrations by aqueous extraction, as an indicator of a high nutritional value for oil extracted.

Other factors that affect the yield of PSO extraction are the pressure and amplitude used. Liu et al. (2012) reported the use of superheated solvent extraction, where a combination of temperature, pressure, and solvents allows rapid and efficient extraction of analytes from several matrices. In ultrasound extraction, the resonant bubble size is proportional to the amplitude of the ultrasonic wave; for that reason, cavitation is caused using high intensities (Goula, 2013).

Some studies reported the use of microwaves as pre-treatment for pomegranate seed. Gaikwad et al. (2017) reported that the use of 60 s and 720 W, as microwave pretreatment before Soxhlet extraction, gave the consequent increase of 1.4-fold oil yield and a reduction of 60 min in extraction time, with respect to control (only Soxhlet). In other work, microwave pretreatment increased oil yield from 27.7% to 34% and from

21.6% to 25.5% compared to those obtained with untreated seeds for Soxhlet and scCO_2 extractions, respectively (Đurđević, 2017). By applying a microwave pretreatment of 6 min at 600 W before a Soxhlet extraction, the authors reported 36% of oil yield, whereas for scCO_2 extraction, the highest oil yield was reached in 6 min at 250 W of microwave pretreatment (27.2%).

All these studies related to PSO extraction, as well as the high amount of PuA present, highlight the importance of this fruit, in particular its byproducts (peel and seeds), as a potential source of value-added biological compounds. These byproducts can be revalorized into interesting molecules for food, cosmetic, and pharmaceutic industries.

6.7 NUTRACEUTICALS BENEFITS

Nutraceuticals are considered dietary ingredients naturally occurring in foods, namely, carotenoids, fatty acids, minerals, polyphenols, and vitamins, with beneficial effects on the health, and they can be used as complementary agents for the treatment of diverse diseases (Khajebishak et al., 2019; Souyoul et al., 2018). The nutritional and antioxidant characteristics of the pomegranate fruit have led to a recent interest in its use as a beneficial source of secondary metabolites because the chemical properties of PSO are significantly superior to those of red wine, green tea, and the synthetic antioxidant butylated hydroxyanisole (Al-Muammar and Khan, 2012; Siano et al., 2015). Extracts from different parts of this plant have been used since ancient times to treat diverse pathologies and nowadays are being investigated for their several pharmacologic activities, such as antibacterial, astringent, antidiarrheal, and antiobesity activities (Zhang et al., 2010).

Recent research has reported that fruit seed oil may function as a health promotor and in disease prevention due to its specific fatty acid composition, showing a nutraceutical effect (Yu et al., 2006). The antioxidant activity of pomegranate constituents has been the subject of many studies conducted *in vivo* and *in vitro*, and this activity is related to the diverse phenolic compounds present in the pomegranate; for example, the higher antioxidant activity of seeds may be explained by the presence of specific phytochemicals such as anthocyanins, tocopherols, and PuA in seed since pomegranate seeds are considered as a natural source of essential fatty acids, particularly PuA (Alcaraz-Mármol et al., 2015; Lansky and Newman, 2007; Orak et al., 2012).

PuA is a polyunsaturated fatty acid, which is classified as a conjugated linolenic acid, that possesses strong antioxidant, antimicrobial, anti-inflammatory, antiatherogenic, anti-invasive, antimetastatic, and antitumorigenic properties (Mete et al., 2019). Some studies have shown the therapeutic effect of this fatty acid for diabetes mellitus management since PuA exerts antidiabetic effects by means of reducing inflammatory cytokines, modulating glucose homeostasis, and antioxidant properties; besides this acid, the other known compounds in pomegranate, such as ellagic, gallic, oleanolic, ursolic, and uallic acids, have been identified for their antidiabetic role (Khajebishak et al., 2019; Banihani et al., 2013).

On the one hand, the protection offered by PSO against diet-induced obesity and insulin resistance is independent of changes in food intake or energy expenditure (Aruna et al., 2016). In a study conducted by Nekooeian et al. (2014), the effects of PSO on rats with type 2 diabetes mellitus were analyzed, and it was observed that rats treated with PSO had significantly higher levels of serum insulin and glutathione peroxidase activity and there were no statistically significant differences in terms of blood glucose between them and the control group. These results suggest that PSO, especially, PuA, improved insulin secretion and decreased plasma glucose. The mechanism of pomegranate oil-induced enhancement in insulin levels is not known, however, this effect could be attributed to diminishing the diabetes-related oxidant stress, reflected in enhanced insulin levels (Bedel et al., 2017). Yamasaki et al. (2006) evaluated the effects of dietary PSO, which contains high levels of PuA, on immune function and lipid metabolism in mice, splenocytes isolated from mice fed with PSO produced larger amounts of immunoglobulins G and M, on the other hand, PSO did not affect the percentages of B cells, which are an essential component of the immune system, analysis of serum lipid parameters showed significant increases in serum triacylglycerol and phospholipid levels but not in total cholesterol. Spilmont et al. (2013) demonstrated that PSO consumption (5% of the diet) in mice, improved significantly bone mineral density, suggesting the involvement of both osteoclastogenesis inhibition, hence, PuA offers promising alternatives to age-related bone complications.

Hence, the consumption of PuA could avoid osteoporosis occurrence in humans since this disease is increasing day by day and gaining importance as a great threat to long and healthy life expectancy (Shabbir et al., 2017). Shaoul et al. (2018) evaluated the effects of oral pomegranate extracts supplementation on intestinal structural changes, enterocyte proliferation, and apoptosis during methotrexate-induced intestinal damage in rats, supplemented rats demonstrated a decrease in enterocyte apoptosis, and this was associated with a decrease in caspase 3 protein expression as well as increased cell proliferation, concluding that with oral pomegranate extracts prevents mucosal injury and improves intestinal recovery following injury in the rat, these observations suggest that immunonutrition containing pomegranate extracts may have clinical utility for human cancer patients treated with cytotoxic drugs to diminish chemotherapy-induced mucositis. On the one hand, pomegranate oil components have been found to inhibit the invasion and proliferation of different cell lines of human cancer since PuA produced the apoptotic effect on cell lines and impaired the cellular mitochondrial membrane potential (Bedel et al., 2017). Yayla et al. (2018) determined the therapeutic effects of PSO, as a powerful antioxidant and anti-inflammatory agent, on ovarian ischemia in rats, finding that low doses of PSO application reduced significantly oxidative stress and NADPH oxidase activity; at the same time, low doses of PSO increased antioxidant activity; hence, PSO demonstrated an important therapeutic effect in the treatment of ovarian ischemia.

Plant oils have the potential to be used for a large number of nutraceutical applications due to bioactive components and can provide a sustainable replacement for current sources of these molecules since favorable effects of these fatty acids on health were observed *in vitro* or in animals; however, more intervention studies, especially on humans, are required to verify these beneficial effects (Li et al., 2018; Singer and Weselake, 2018).

6.8 FOOD APPLICATIONS

The notion that food may possess the ability to prevent disease and be used as a treatment of ailments is ancient. Nowadays, there has been an increased interest in food containing a high amount of polyunsaturated fatty acids because these fatty acids are considered as functional ingredients to prevent chronic diseases, recent investigations have focused on naturally occurring molecules to satisfy consumer concerns over safety and toxicity of food additives (Emami et al., 2015; Faria and Calhau, 2010). The more common sources of these functional specialty oils include marine organisms, tree nuts, cereals, and berry plants, these fatty acids are used in nutraceutical foods products mainly for their higher amounts of monounsaturated fatty acids (Hernandez, 2016). Fruit and vegetable transformation produces a huge amount of byproducts that are considered to have a lot of bioactive compounds, for example, the pomegranate transformation to various edible products mainly nectars, juices, jams, and jellies; however, during this transformation procedure, a large quantity of wastes is produced, which has lots of nutritional components but most of them are wasted, causing environmental pollution and the loss of valuable nutritional components, such as fatty acids present in the seeds since seeds represent the portion of the fruit with the highest concentration of bioactive molecules. In pomegranate fruits, bioactive molecules are often abundant, although their amount depends on the cultivar, geographical location, growing conditions, and maturity stage, so that their waste represents a double loss for agrifood industry that has to face the cost of disposal and the loss of profits for their reuse and valorization. In this sense, in recent years, seed oils have been receiving interest of the food industry due to their high concentrations of hydrophilic and lipophilic bioactive components, which have important pharmacological properties on human health (Alcaraz-Mármol et al., 2015; Durante et al., 2017; Shabbir et al., 2017; Siano et al., 2015). PSO is characterized by its fatty acid composition, of which approximately 80% is PuA; this acid is believed to exert several physiological functions as an active biological component. PuA was granted GRAS status by the FDA and is actively sold commercially as a weight control and antiobesity supplement in health stores; however, no health claims have been permitted (Hernandez, 2016; Koba and Yanagita, 2011). Some studies have been conducted to assess PuA safety, and it has not been reported to possess deleterious health effects by its supplementation (Faria and Calhau, 2010). Despite this, PuA has not been extensively explored as an ingredient of food products that can be aimed at specific consumer target groups (Aruna et al., 2016). For example, Çam et al. (2013) evaluated the incorporation of pomegranate peel phenolics and PSO to ice cream because it is poor in poly unsaturated fatty acids and phenolics, resulting in significant changes in the pH, total acidity, and color of the samples. Besides the enrichment of ice creams with pomegranate byproducts, these products might provide health benefits to consumers with functional properties of punicalagins in pomegranate peel and PuA in PSO. Goula and Adamopoulos (2012) proposed the encapsulation by spray drying of pomegranate seeds to the application in food industries to prevent oxidative degradation that results in a loss of nutritional quality and development of undesired flavors of oils.

Modaresi et al. (2011) and Emami et al. (2016) evaluated the effects of feeding goats with pomegranate seed pulp on milk yield, milk composition, and fatty acid profiles of milk fat, showing that the milk fat concentration of goats fed with pomegranate seed pulp diets increased but milk yield, milk protein, and milk solids-not-fat concentration were not affected by diets, concluding that feeding goats with pomegranate seed pulp modified the milk fatty acid profile, including conjugated linoleic, punicic, and vaccenic acids, without negative effects on the intake, milk yield, and nutrient digestibility. Kostogrys et al. (2017) determined the effects of PSO, used as a source of PuA in the diets of laying hens, on the physicochemical properties of eggs and observing that eggs naturally enriched with PuA preserve their composition and conventional properties (chemical composition and physical and organoleptic properties); dietary PuA had a positive impact on the color of the egg yolk, whereas the hardness of egg yolks was not affected. Additionally, increasing dietary PuA led to an increase of PuA in egg-yolk lipids. Banaszkiewicz et al. (2018) evaluated the value and quality after the slaughter of broiler chicken meat fed with PSO, and the possibility of enriching it in fatty acids, and showed that PSO significantly improved the palatability of thigh muscles and influenced the fatty acid profile of the meat; the inclusion of PSO resulted in the deposition of a small amount of PuA while significantly increasing rumenic acid and improving the quality of broiler chicken meat. Acar et al. (2018) assessed the effects of PSO diet on growth performance, some hematological, biochemical, and immunological parameters, and disease resistance against *Yersinia ruckeri* in cultured rainbow trout *Oncorhynchus mykiss* and an increase weight, growth, and feed conversion were found in fish fed with PSO diets after the feeding trial; besides, dietary administration of PSO induced a reduction in mortality of rainbow trout infected with *Y. ruckeri*. The edible oil found in pomegranates is quite rare; oil components contribute large amounts to the nutritional and sensory properties, besides presenting a strong antioxidant effect, attributed mainly to PuA; due to this, the consumption of pomegranate and its derivatives, such as the oil seeds, has increased dramatically worldwide, and hence, gathering information about the composition of PuA, its components, and its functional properties is essential to support the development of new products with potential applications in the food industry, providing an important value nutritional (Alcaraz-Mármol et al., 2015; Faria, 2010; Loizzo et al., 2019).

6.9 SCIENTIFIC PERSPECTIVES

Lipids are important in human health. Besides, consumers are more concerned about their nutrition. It has been described in the present chapter that residues from pomegranate processing are suitable for the extraction of bioactive compounds. In particular, PuA is highlighted for its important bioactivities. Then, future research can be conducted for the synthesis of structured lipids (glycerides) with elevated PuA content by means of enzyme-catalyzed reactions. On the other hand, the formulation of nano-emulsions containing PuA can act as a transporter of important metabolites. Furthermore, the food (bread, cookies, snacks) can be prepared using a high-PuA oil. The different applications imply the evaluation of reaction system, omics tools, stability, and storage of the products obtained and their feasibility.

KEYWORDS

- **pomegranate seeds**
- **fatty acid**
- **pharmaceutical applications**
- **nutraceutical properties**
- **bioactivities**

REFERENCES

Abbasi, H.; Rezaei, K.; Rashidi, L., Extraction of essential oils from the seeds of pomegranate using organic solvents and supercritical CO_2. *J. Am. Oil Chem. Soc.* 2008, *85* (1), 83–89.

Acar, Ü.; Parrino, V.; Kesbiç, O. S.; Lo Paro, G.; Saoca, C.; Abbate, F.; Yılmaz, S.; Fazio, F., Effects of different levels of pomegranate seed oil on some blood parameters and disease resistance against *Yersinia ruckeri* in rainbow trout. *Front. Physiol.* 2018, *9* (596). DOI: 10.3389/fphys.2018.00596.

Ahangari, B.; Sargolzaei, J., Extraction of pomegranate seed oil using subcritical propane and supercritical carbon dioxide. *Theor. Found. Chem. Eng.*2012, *46* (3), 258–265.

Akhtar, S.; Ismail, T.; Fraternale, D.; Sestili, P., Pomegranate peel and peel extracts: chemistry and food features. *Food Chem.* 2015, *174*, 417–425.

Alcaraz-Mármol, F.; Nuncio-Jáuregui, N.; Calín-Sánchez, Á.; Carbonell-Barrachina, Á. A.; Martínez, J. J.; Hernández, F., Determination of fatty acid composition in arils of 20 pomegranates cultivars grown in Spain. *Sci. Hort.* 2015,*197*, 712–718.

Al-Muammar, M. N.; Khan, F., Obesity: The preventive role of the pomegranate (*Punica granatum*). *Nutrition* 2012, *28* (6), 595–604.

Amri, Z.; Zaouay, F.; Lazreg-Aref, H.; Soltana, H.; Mneri, A.; Mars, M.; Hammami, M., Phytochemical content, Fatty acids composition and antioxidant potential of different pomegranate parts: Comparison between edible and non edible varieties grown in Tunisia. *Int. J. Biol. Macromol.* 2017, *104*, 274–280.

Aruna, P.; Manohar, B.; Singh, R. P., Processing of pomegranate seed waste and mass transfer studies of extraction of pomegranate seed oil. *J. Food Proces. Preserv.* 2018, *42* (5), e13609.

Aruna, P.; Venkataramanamma, D.; Singh, A. K.; Singh, R., Health benefits of punicic acid: a review. *Compr. Rev. Food Sci. Food Saf.* 2016, *15* (1), 16–27.

Ascacio-Valdés, J. A.; Buenrostro-Figueroa, J. J.; Aguilera-Carbó, A.; Prado-Barragán, A.; Rodríguez-Herrera, R.; Aguilar, C. N., Ellagitannins: biosynthesis, biodegradation and biological properties *J. Med. Plants. Res.* 2011, *5* (19), 4696–4703.

Banaszkiewicz, T.; Białek, A.; Tokarz, A.; Kaszperuk, K., Effect of dietary grape and pomegranate seed oil on the post-slaughter value and physicochemical properties of muscles of broiler chickens. *Acta Sci. Pol. Technol. Aliment.* 2018, *17* (3), 199–209.

Banihani, S.; Swedan, S.; Alguraan, Z., Pomegranate and type 2 diabetes. *Nutr. Res.* 2013, *33* (5), 341–348.

Barizão, É. O.; Boeing, J. S.; Martins, A. C.; Visentainer, J. V.; Almeida, V. C., Application of response surface methodology for the optimization of ultrasound-assisted extraction of pomegranate (*Punica granatum* L.) seed oil. *Food Anal. Method.* 2015, *8* (9), 2392–2400.

Bassiri-Jahromi, S., *Punica granatum* (Pomegranate) activity in health promotion and cancer prevention. *Oncol. Rev.* 2018,.*12* (1), 345.

Bedel, H. A.; Turgut, N. T.; Kurtoglu, A. U.; Usta, C., Effects of nutraceutical punicic acid. *Indian J. Pharm. Sci.* 2017, *79* (3), 328–334.

Białek, A.; Białek, M.; Lepionka, T.; Kaszperuk, K.; Banaszkiewicz, T.; Tokarz, A., The effect of pomegranate seed oil and grapeseed oil on *cis*-9, *trans*-11 CLA (rumenic acid), n-3 and n-6 fatty acids deposition in selected tissues of chickens. *J. Anim. Physiol. Anim. Nutr.* 2018, *102* (4), 962–976.

Calani, L.; Beghè, D.; Mena, P.; Del Rio, D.; Bruni, R.; Fabbri, A.; Dall'Asta, C.; Galaverna, G., Ultra-HPLC–MSn (poly)phenolic profiling and chemometric analysis of juices from ancient *Punica granatum* L. cultivars: a nontargeted approach. *J. Agric. Food Chem.* 2013, *61* (23), 5600–5609.

Çam, M.; Erdoğan, F.; Aslan, D.; Dinç, M., Enrichment of functional properties of ice cream with pomegranate by-products. *J. Food Sci.* 2013, *78* (10), C1543–C1550.

Çam, M.; Hışıl, Y., Pressurised water extraction of polyphenols from pomegranate peels. *Food Chem.* 2010, *123* (3), 878–885.

Çam, M.; İçyer, N. C., Phenolics of pomegranate peels: extraction optimization by central composite design and alpha glucosidase inhibition potentials. *J. Food Sci. Technol.* 2015, *52* (3), 1489–1497.

Çavdar, H. K.; Yanık, D. K.; Gök, U.; Göğüş, F., Optimisation of microwave-assisted extraction of pomegranate (*Punica granatum* L.) seed oil and evaluation of its physico-chemical and bioactive properties. *Food Technol. Biotechnol.* 2017, *55* (1), 86–94.

Costa, A. M. M.; Silva, L. O.; Torres, A. G., Chemical composition of commercial cold-pressed pomegranate (*Punica granatum*) seed oil from Turkey and Israel, and the use of bioactive compounds for samples' origin preliminary discrimination. *J. Food Compos. Anal.* 2019, *75*, 8–16.

dos Santos Souza, A.; de Souza Jr., J. R.; Sousa, D. C. P.; Albuquerque, U. P., *Punica granatum* L. In *Medicinal and Aromatic Plants of South America: Brazil*, Albuquerque, U. P.; Patil, U.; Máthé, Á., Eds. Springer Netherlands: Dordrecht, 2018; pp. 413–420.

Durante, M.; Montefusco, A.; Marrese, P. P.; Soccio, M.; Pastore, D.; Piro, G.; Mita, G.; Lenucci, M. S., Seeds of pomegranate, tomato and grapes: an underestimated source of natural bioactive molecules and antioxidants from agri-food by-products. *J. Food Compos. Anal.* 2017, *63*, 65–72.

Đurđević, S., Improvement of supercritical CO2 and n-hexane extraction of wild growing pomegranate seed oil by microwave pretreatment. *Ind. Crops Prod.* 2017, *104*, 21–27.

Đurđević, S.; Šavikin, K.; Živković, J.; Böhm, V.; Stanojković, T.; Damjanović, A.; Petrović, S., Antioxidant and cytotoxic activity of fatty oil isolated by supercritical fluid extraction from microwave pretreated seeds of wild growing *Punica granatum* L. *J. Supercrit. Fluid* 2018, *133*, 225–232.

Eikani, M. H.; Golmohammad, F.; Homami, S. S., Extraction of pomegranate (*Punica granatum* L.) seed oil using superheated hexane. *Food Bioprod. Process.* 2012, *90* (1), 32–36.

Emami, A.; Fathi Nasri, M. H.; Ganjkhanlou, M.; Rashidi, L.; Zali, A., Dietary pomegranate seed pulp increases conjugated-linoleic and -linolenic acids in muscle and adipose tissues of kid. *Anim. Feed Sci. Technol.* 2015, *209*, 79–89.

Emami, A.; Fathi Nasri, M. H.; Ganjkhanlou, M.; Rashidi, L.; Zali, A., Effect of pomegranate seed oil as a source of conjugated linolenic acid on performance and milk fatty acid profile of dairy goats. *Livest. Sci.* 2016, *193*, 1–7.

Erkan, M.; Kader, A. A., 14 - Pomegranate (*Punica granatum* L.) A2 Yahia, Elhadi M. In *Postharvest Biology and Technology of Tropical and Subtropical Fruits*, Woodhead Publishing: Cambridge, UK, 2011; pp. 287e –313e.

FAO, Statistical database. Food and Agriculture Organization of the United Nations, 2014.

Faria, A.; Calhau, C., Chapter 36 - Pomegranate in human health: an overview. In *Bioactive Foods in Promoting Health*, Watson, R. R.; Preedy, V. R., Eds. Academic Press: San Diego, 2010; pp. 551–563.

Fischer, U. A.; Carle, R.; Kammerer, D. R., Identification and quantification of phenolic compounds from pomegranate (*Punica granatum* L.) peel, mesocarp, aril and differently produced juices by HPLC-DAD–ESI/MSn. *Food Chem.* 2011, *127* (2), 807–821.

Fischer, U. A.; Jaksch, A. V.; Carle, R.; Kammerer, D. R., Influence of origin source, different fruit tissue and juice extraction methods on anthocyanin, phenolic acid, hydrolysable tannin and isolariciresinol contents of pomegranate (*Punica granatum* L.) fruits and juices. *Eur. Food Res. Technol.* 2013, *237* (2), 209–221.

Gaikwad, N. N.; Yedle, V. H.; Yenge, G.; Suryavanshi, S., Effect of microwave pretreatment on extraction yield of pomegranate seed (cv. Bhagwa) oil. *Int. J. Chem. Stud.* 2017, *5* (4), 1292–1295.

Ghorbanzadeh, R.; Rezaei, K., Optimization of an aqueous extraction process for pomegranate seed oil. *J. Am. Oil Chem. Soc.*2017, *94* (12), 1491–1501.

Gil, M. I.; Tomás-Barberán, F. A.; Hess-Pierce, B.; Holcroft, D. M.; Kader, A. A., Antioxidant activity of pomegranate juice and its relationship with phenolic composition and processing. *J. Agric. Food Chem.* 2000, *48* (10), 4581–4589.

Goula, A. M., Ultrasound-assisted extraction of pomegranate seed oil – kinetic modeling. *J. Food Eng.* 2013, *117* (4), 492–498.

Goula, A. M.; Adamopoulos, K. G., A method for pomegranate seed application in food industries: seed oil encapsulation. *Food Bioprod. Process.* 2012, *90* (4), 639–652.

Hernandez, E. M., Specialty oils: functional and nutraceutical properties. In *Functional Dietary Lipids*, Sanders, T., Ed. Woodhead Publishing Elsevier: London, UK, 2016; pp. 69–101.

Holic, R.; Xu, Y.; Caldo, K. M. P.; Singer, S. D.; Field, C. J.; Weselake, R. J.; Chen, G., Bioactivity and biotechnological production of punicic acid. *Appl. Microbiol. Biotechnol.* 2018, *102* (8), 3537–3549.

Hornung, E.; Pernstich, C.; Feussner, I., Formation of conjugated $\Delta^{11}\Delta^{13}$-double bonds by Δ^{12}-linoleic acid (1,4)-acyl-lipid-desaturase in pomegranate seeds. *Eur. J. Biochem.* 2002, *269* (19), 4852–4859.

Horuz, E.; Maskan, M., Hot air and microwave drying of pomegranate (*Punica granatum* L.) arils. *J. Food Sci. Technol.* 2015, *52* (1), 285–293.

Iwabuchi, M.; Kohno-Murase, J.; Imamura, J., Δ^{12}-oleate desaturase-related enzymes associated with formation of conjugated trans-Δ^{11}, cis-Δ^{13} double bonds. *J. Biol. Chem.* 2003, *278* (7), 4603–4610.

Karimi, M.; Sadeghi, R.; Kokini, J., Pomegranate as a promising opportunity in medicine and nanotechnology. *Trends Food Sci. Technol.* 2017,*69*, 59–73.

Kaufman, M.; Wiesman, Z., Pomegranate oil Analysis with emphasis on MALDI-TOF/MS triacylglycerol fingerprinting. *J. Agric. Food Chem.* 2007, *55* (25), 10405–10413.

Khajebishak, Y.; Payahoo, L.; Alivand, M.; Alipour, B., Punicic acid: a potential compound of pomegranate seed oil in Type 2 diabetes mellitus management. *J. Cell. Physiol.* 2019, *234* (3), 2112–2120.

Khoddami, A.; Man, Y. B. C.; Roberts, T. H., Physico-chemical properties and fatty acid profile of seed oils from pomegranate *(Punica granatum* L.) extracted by cold pressing. *Eur. J. Lipid Sci. Technol.* 2014, *116* (5), 553–562.

Koba, K.; Yanagita, T., Chapter 108—Potential health benefits of pomegranate (*Punica granatum*) seed oil containing conjugated linolenic acid. In *Nuts and Seeds in Health and Disease Prevention*, Preedy, V. R.; Watson, R. R.; Patel, V. B., Eds. Academic Press: San Diego, 2011; pp. 919–924.

Kostogrys, R. B.; Filipiak-Florkiewicz, A.; Dereń, K.; Drahun, A.; Czyżyńska-Cichoń, I.; Cieślik, E.; Szymczyk, B.; Franczyk-Żarów, M., Effect of dietary pomegranate seed oil on laying hen performance and physicochemical properties of eggs. *Food Chem.* 2017, *221*, 1096–1103.

Kýralan, M.; Gölükcü, M.; Tokgöz, H., Oil and conjugated linolenic acid contents of seeds from important pomegranate cultivars (*Punica granatum* L.) grown in Turkey. *J. Am. Oil Chem. Soc.* 2009, *86* (10), 985–990.

Lansky, E. P.; Newman, R. A., *Punica granatum* (pomegranate) and its potential for prevention and treatment of inflammation and cancer. *J. Ethnopharmacol.* 2007, *109* (2), 177–206.

Li, K.; Sinclair, A.; Zhao, F.; Li, D., Uncommon fatty acids and cardiometabolic health. *Nutrients* 2018, *10* (10), 1559.

Liu, G.; Xu, X.; Gong, Y.; He, L.; Gao, Y., Effects of supercritical CO_2 extraction parameters on chemical composition and free radical-scavenging activity of pomegranate (Punica granatum L.) seed oil. *Food Bioprod. Process.* 2012, *90* (3), 573–578.

Loizzo, M. R.; Aiello, F.; Tenuta, M. C.; Leporini, M.; Falco, T.; Tundis, R., Chapter 3.46 - Pomegranate (*Punica granatum* L.). In *Nonvitamin and Nonmineral Nutritional Supplements*, Nabavi, S. M.; Silva, A. S., Eds. Academic Press: San Diego, USA, 2019; pp. 467–472.

Meerts, I. A. T. M.; Verspeek-Rip, C. M.; Buskens, C. A. F.; Keizer, H. G.; Bassaganya-Riera, J.; Jouni, Z. E.; van Huygevoort, A. H. B. M.; van Otterdijk, F. M.; van de Waart, E. J., Toxicological evaluation of pomegranate seed oil. *Food Chem. Toxicol.* 2009, *47* (6), 1085–1092.

Mercado Silva, E.; Mondragón Jacobo, C.; Rocha Peralta, L.; Álvarez Mayorga, B., Efectos de condición del fruto y temperatura de almacenamiento en la calidad de granada roja. *Rev. Mex. De Cienc. Agric.* 2011, *2*, 449–459.

Mete, M.; Unsal, U.; Aydemir, I.; Sonmez, P.; Tuglu, M., Punicic acid inhibits glioblastoma migration and proliferation via the PI3K/AKT1/mTOR signaling pathway. *Anticancer Agents Med. Chem.* 2019, 19(19), 1120–1131.

Miranda, J.; Aguirre, L.; Fernández-Quintela, A.; Macarulla, M. T.; Martínez-Castaño, M. G.; Ayo, J.; Bilbao, E.; Portillo, M. P., Effects of pomegranate seed oil on glucose and lipid metabolism-related organs in rats fed an obesogenic diet. *J. Agric. Food Chem.* 2013, *61* (21), 5089–5096.

Modaresi, J.; Fathi Nasri, M. H.; Rashidi, L.; Dayani, O.; Kebreab, E., Short communication: effects of supplementation with pomegranate seed pulp on concentrations of conjugated linoleic acid and punicic acid in goat milk. *J. Dairy Sci.* 2011, *94* (8), 4075–4080.

Nekooeian, A. A.; Eftekhari, M. H.; Adibi, S.; Rajaeifard, A., Effects of pomegranate seed oil on insulin release in rats with type 2 diabetes. *Iran. J. Med. Sci.* 2014, *39* (2), 130.

Nuncio-Jáuregui, N.; Calín-Sánchez, Á.; Vázquez-Araújo, L.; Pérez-López, A. J.; Frutos-Fernández, M. J.; Carbonell-Barrachina, Á. A., Processing pomegranates for juice and impact on bioactive components A2. In *Processing and Impact on Active Components in Food*, Preedy, V. Ed., Academic Press: San Diego, 2015; pp. 629–636; chap. 76.

Orak, H. H.; Yagar, H.; Isbilir, S. S., Comparison of antioxidant activities of juice, peel, and seed of pomegranate (*Punica granatum* L.) and inter-relationships with total phenolic, tannin, anthocyanin, and flavonoid contents. *Food Sci. Biotechnol.* 2012, *21* (2), 373–387.

Özgül-Yücel, S., Determination of conjugated linolenic acid content of selected oil seeds grown in Turkey. *J. Am. Oil Chem. Soc.* 2005, *82* (12), 893–897.

Sassano, G.; Sanderson, P.; Franx, J.; Groot, P.; van Straalen, J.; Bassaganya-Riera, J., Analysis of pomegranate seed oil for the presence of jacaric acid. *J. Sci. Food Agric.* 2009, *89* (6), 1046–1052.

Schneider, A.-C.; Beguin, P.; Bourez, S.; Perfield, J. W., II; Mignolet, E.; Debier, C.; Schneider, Y.-J.; Larondelle, Y., Conversion of t11t13 CLA into c9t11 CLA in Caco-2 cells and inhibition by sterculic oil. *PLoS One* 2012, *7* (3), e32824.

Schneider, A.-C.; Mignolet, E.; Schneider, Y.-J.; Larondelle, Y., Uptake of conjugated linolenic acids and conversion to cis-9, trans-11-or trans-9, trans-11-conjugated linoleic acids in Caco-2 cells. *British J. Nutr.* 2012, *109* (1), 57–64.

Seeram, N. P.; Adams, L. S.; Henning, S. M.; Niu, Y.; Zhang, Y.; Nair, M. G.; Heber, D., In vitro antiproliferative, apoptotic and antioxidant activities of punicalagin, ellagic acid and a total pomegranate tannin extract are enhanced in combination with other polyphenols as found in pomegranate juice. *J. Nutr. Biochem.* 2005, *16* (6), 360–367.

Sepúlveda, L.; Aguilera-Carbó, A.; Ascacio-Valdés, J. A.; Rodríguez-Herrera, R.; Martínez-Hernández, J. L.; Aguilar, C. N., Optimization of ellagic acid accumulation by *Aspergillus niger* GH1 in solid state culture using pomegranate shell powder as a support. *Proc. Biochem.* 2012, *47* (12), 2199–2203.

Shaban, N. Z.; El-Kersh, M. A. L.; El-Rashidy, F. H.; Habashy, N. H., Protective role of *Punica granatum* (pomegranate) peel and seed oil extracts on diethylnitrosamine and phenobarbital-induced hepatic injury in male rats. *Food Chem.* 2013, *141* (3), 1587–1596.

Shabbir, M. A.; Khan, M. R.; Saeed, M.; Pasha, I.; Khalil, A. A.; Siraj, N., Punicic acid: A striking health substance to combat metabolic syndromes in humans. *Lipids Health Dis.* 2017, *16* (1), 99.

Shaoul, R.; Moati, D.; Schwartz, B.; Pollak, Y.; Sukhotnik, I., Effect of pomegranate juice on intestinal recovery following methotrexate-induced intestinal damage in a rat model. *J. Am. Coll. Nutr.* 2018, *37* (5), 406–414.

Siano, F.; Straccia, M. C.; Paolucci, M.; Fasulo, G.; Boscaino, F.; Volpe, M. G., Physico-chemical properties and fatty acid composition of pomegranate, cherry and pumpkin seed oils. *J. Sci. Food Agric.* 2015, *96* (5), 1730–1735.

Singer, S. D.; Weselake, R. J., Production of other bioproducts from plant oils. In *Plant bioproducts*, Chen, G.; Weselake, R. J.; Singer, S. D., Eds. Springer: New York, USA, 2018; pp. 59–85.

Singh, R. P.; Chidambara Murthy, K. N.; Jayaprakasha, G. K., Studies on the antioxidant activity of pomegranate (*Punica granatum*) peel and seed extracts using in vitro models. *J. Agric. Food Chem.* 2002, *50* (1), 81–86.

Soetjipto, H.; Pradipta, M.; Timotius, K., Fatty acids composition of red and purple pomegranate (*Punica granatum* L) seed oil. *Indonesian J. Cancer Chemoprev.* 2010, *1* (2), 74–77.

Souyoul, S. A.; Saussy, K. P.; Lupo, M. P., Nutraceuticals: a review. *Dermatol. Ther.* 2018, *8* (1), 5–16.

Spilmont, M.; Léotoing, L.; Davicco, M.-J.; Lebecque, P.; Mercier, S.; Miot-Noirault, E.; Pilet, P.; Rios, L.; Wittrant, Y.; Coxam, V., Pomegranate seed oil prevents bone loss in a mice model of osteoporosis, through osteoblastic stimulation, osteoclastic inhibition and decreased inflammatory status. *J. Nutr. Biochem.* 2013, *24* (11), 1840–1848.

Szychowski, P. J.; Frutos, M. J.; Burló, F.; Pérez-López, A. J.; Carbonell-Barrachina, Á. A.; Hernández, F., Instrumental and sensory texture attributes of pomegranate arils and seeds as affected by cultivar. *LWT—Food Sci. Technol.* 2015, *60* (2, Part 1), 656–663.

Tian, Y.; Xu, Z.; Zheng, B.; Martin Lo, Y., Optimization of ultrasonic-assisted extraction of pomegranate (*Punica granatum* L.) seed oil. *Ultrason. Sonochem.* 2013, *20* (1), 202–208.

Topkafa, M.; Kara, H.; Sherazi, S. T. H., Evaluation of the triglyceride composition of pomegranate seed oil by RP-HPLC followed by GC-MS. *J. Am. Oil Chem. Soc.* 2015, *92* (6), 791–800.

Tsuzuki, T.; Kawakami, Y.; Abe, R.; Nakagawa, K.; Koba, K.; Imamura, J.; Iwata, T.; Ikeda, I.; Miyazawa, T., Conjugated linolenic acid is slowly absorbed in rat intestine, but quickly converted to conjugated linoleic acid. *J. Nutr.* 2006, *136* (8), 2153–2159.

USDA, National Nutrient Database for Standard Reference. United States Department of Agriculture, 2016. Accessed 05/05/2019.

Viuda-Martos, M.; Fernández-López, J.; Pérez-Álvarez, J. A., Pomegranate and its many functional components as related to human health: a review. *Compr. Rev. Food Sci. Food Saf.* 2010, *9* (6), 635–654.

Vroegrijk, I. O. C. M.; van Diepen, J. A.; van den Berg, S.; Westbroek, I.; Keizer, H.; Gambelli, L.; Hontecillas, R.; Bassaganya-Riera, J.; Zondag, G. C. M.; Romijn, J. A.; Havekes, L. M.; Voshol, P. J., Pomegranate seed oil, a rich source of punicic acid, prevents diet-induced obesity and insulin resistance in mice. *Food Chem. Toxicol.* 2011, *49* (6), 1426–1430.

Wasila, H.; Li, X.; Liu, L.; Ahmad, I.; Ahmad, S., Peel effects on phenolic composition, antioxidant activity, and making of pomegranate juice and wine. *J. Food Sci.* 2013, *78* (8), C1166–C1172.

Yamasaki, M.; Kitagawa, T.; Koyanagi, N.; Chujo, H.; Maeda, H.; Kohno-Murase, J.; Imamura, J.; Tachibana, H.; Yamada, K., Dietary effect of pomegranate seed oil on immune function and lipid metabolism in mice. *Nutrition* 2006, *22* (1), 54–59.

Yayla, M.; Cetin, D.; Adali, Y.; Aksu Kilicle, P.; Toktay, E., Potential therapeutic effect of pomegranate seed oil on ovarian ischemia/reperfusion injury in rats. *Iran. J. Basic Med. Sci.* 2018, *21* (12), 1262–1268.

Yu, L.; Parry W., J.; Zhou, K., Fruit Seed Oils. In *Nutraceutical and Specialty Lipids and Their Co-products*, Shaidi, F., Ed. CRC Press: Florida, USA, 2006; p. 584.

Yuan, G.; Sinclair, A. J.; Xu, C.; Li, D., Incorporation and metabolism of punicic acid in healthy young humans. *Mol. Nutr. Food Res.*2009, *53* (10), 1336–1342.

Yuan, G.-F.; Yuan, J.-Q.; Li, D., Punicic acid from trichosanthes kirilowii seed oil is rapidly metabolized to conjugated linoleic acid in rats. *J. Med. Food* 2009, *12* (2), 416-422.

Zhang, L.; Gao, Y.; Zhang, Y.; Liu, J.; Yu, J., Changes in bioactive compounds and antioxidant activities in pomegranate leaves. *Sci. Hortic.* 2010, *123* (4), 543–546.

Zhu, C.; Liu, X., Optimization of extraction process of crude polysaccharides from pomegranate peel by response surface methodology. *Carbohydr. Polym.* 2013, *92* (2), 1197–1202.

CHAPTER 7

NUTRITIONAL EVALUATION OF WASTE IN THE CITRUS INDUSTRY

PATRICIA M. ALBARRACÍN*, MARÍA F. LENCINA, and NORMA G. BARNES

Process Engineering and Infdustrial Management Department, Faculty of Exact Sciences and Technology, National University of Tucumán, Argentina

**Corresponding author. E-mail: palbarracin@herrera.unt.edu.ar*

ABSTRACT

In the citrus industry, as byproducts of the process to obtain citrus juices, there remain the crusts, membranes, part of the pulp, and seeds. This chapter shows the nutritional study of wet husk, husk dust, residue from the crushing of the dry skin, and wet residue from the pulp and solids filtered from the liquid effluents of the plant. This study was carried out with the aim to analyze the use of these wastes in the elaboration of balanced feed for animals. Samples were taken from a local citrus industry. The contents of dry matter, lipids, fiber, protein,s carbohydrates, and flavonoids were determined in triplicate. The obtained values were analyzed statistically in Excel 2010 and showed a standard deviation on the order of 5%. The powder of lemon peel showed the highest content of dry matter (99.7%), followed by lipids, proteins, fibers, and carbohydrates. For the wet shell, the content was 37.7%, and for the wet residues filtered, the content was 23.5%. The obtained results allow affirming that the industrial waste analyzed can be used in animal balanced feed, making supplementation of the nutrients in defect.

7.1 INTRODUCTION

The processing of citrus is different from that of other fruits; juice, peel, and essential oils must be separated as perfectly as possible, while it is neither convenient the complete trituration of the fruits nor the obtention of pasta through pressing. As subproducts of the citric juice, the peel, membranes, part of the pulp, and eventually the seeds are left. These are employed as forage, with 100% destined to overseas markets, mainly Denmark and Holland (Federcitrus, 2010). The dehydrated peel of the citrus is used in the province of Misiones as a complement in the feeding of cattle in formulas for balanced food (Coppo and Mussart de Coppo, 2006).

The dried lemon peel is obtained at the end of the process, after the extraction of juice and essential oils; it is dried and triturated to be sold for extracting pectins. From the crasher comes out some dust as a waste. The wet waste of the pulp is the result of the filtering of liquid effluents from the plant where the liquid is used for watering.

Due to the great amount of fiber, the lemon peel could be transformed into a nutraceutical, an ailment that is beneficial to our health. These foods are often called "functional ailments," being the lemon peel appropriate for the elaboration of a functional veterinary food.

The nutritional value of the food depends to a great extent on the quality of the ingredients used, which are measured according to their capacity of promoting growth and maintaining organisms in good states. Thus, the origin or the previous treatment to which the materials are exposed will influence their quality, depending on the availability of nutrients to be used in the metabolic functions of the animal, or the content of endogenous toxic characteristic of each material, particularly those of vegetal origin, which function as antinutrients affecting the health of the organism.

In this chapter, we will present the results of a nutritional study of the waste of the citrus industry, in peel dust, wet peel, and waste of lemon, as we have determined dried mater, lipids, ailment fibers, proteins, and hydrates of carbon, with the purpose of analyzing the possibility of the use in the same elaboration of the balanced food for animals.

7.2 MATERIALS AND METHODS

These were analyzed using the waste of lemon (solids filtered from the final liquid effluents), wet peel, and dust peel (waste resulting from the drying of the peel). These were taken from the process of a local citric factory. Six samples were analyzed from each waste, and three determinations of each variable were analyzed. For the statistical analysis of the data, an Excel 2010 calculus sheet was used, with standard deviation being employed to measure the dispersion of the values with respect to the medium.

The dried matter (DM), lipids (L), alimentary fiber (F), percentage of raw proteins (P), and carbohydrates (CH) were determined. The parameters were studied using the Association of Official Agricultural Chemists (AOAC) techniques (Latimer, 2012).

The mass spectrometry MS was determined through the gravimetric method in a drying oven at 105 °C on samples of 7–8 g of bulk, using a Mettler analytical balance with an appreciation of 0.01 g.

The proteins were determined through the Kjeldahl method over samples of 0.5–1.0 g precisely weighted after proceeding with the digestion, distillation, and identification, as indicated in the technique (Garcia Garrido and Rodriguez Lopez, 2004). The data obtained are numbers of soulish N_2 from each sample in the experiment and the percentage corresponding to proteins, which are calculated on the basis of N_2 product multiplied by the factor of 5.71 (National Institution of Nutrition, 1994).

The accuracy of the determination was on the order of 10^{-2} g.

For the determination of lipids, the Soxhlet method for direct extraction was used. The soluble material in ether was informed as raw fat or ether extract with an accurateness of 10^{-2} g.

The alimentary fiber was determined by digestion of the sample with sodium hydroxide and sulfuric acid. After the incineration in muffle oven during 30 min, through weighing, it was determined in total unsolvable matter constituted by alimentary fibers and ashes. The alimentary fiber was the result of subtracting these values of the ashes. The accurateness of the measure was on the order of 10^{-3} g.

The ashes (A) were determined through the calcination of the samples on a muffle oven at 550 °C for 12 h. These are considered as the content of total minerals or inorganic matter in the sample.

The carbohydrates (CH) were calculated through difference, using the following equation:

$$CH = \text{Total matter} - (W+P+L+F) \tag{7.1}$$

where W is water content = Total matter − DM.

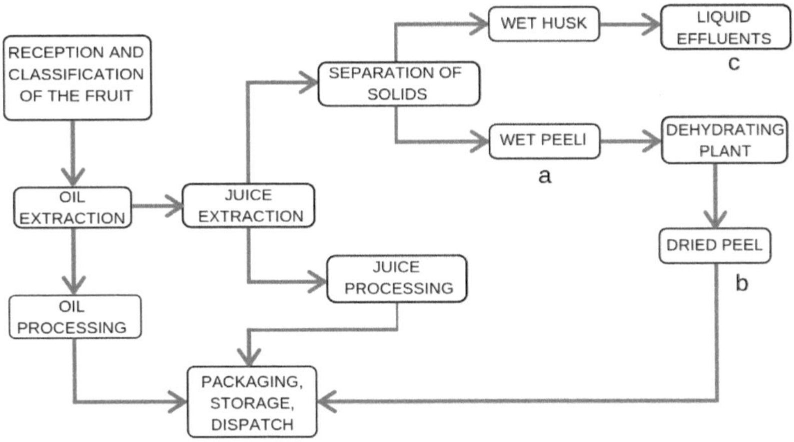

FIGURE 7.1 Graph of a citrus processing factory and the identification of the sectors of our sample extractions (a–c).

This data was compared to the contents of another animal balanced food used as a control, for example, the Citrusvil diet, which contained 25% of gluten feed, 37% of mineral premix, and 0.65% of urea (Lezcano, 2010).

7.3 RESULTS AND DISCUSSION

Table 7.1 shows the percentage of dried matter for the different industrial wastes analyzed and for animal food used as a control. Cisint et al. (2007) propose the lemon peel dust as possible balanced food concentrated by its high content of dried matter. The low content of water in the sample of dust would reduce the risk of microbial contamination, making more vulnerable in this process the samples of wet peel and lemon waste.

TABLE 7.1 Values in Percentages of DM in the Waste of the Citric Factory and in a Control Animal Food

Parameters ($X \pm \delta$)	Lemon Waste	Wet Peel	Dust Peel	Witness Animal Food
%DM	23.46 ± 0.23	37.70 ± 0.36	99.80 ± 2.10	64.47 ± 2.25

Note: *X*, average value of the experiments realized; *δ*, standard deviation.

Table 7.2 shows the values of proteins, lipids, fibers, and carbohydrates in citric waste and in the control animal food obtained in each sample.

The nutritional value of a protein is directly related to its composition, quantity, and proportion of essential amino acids (Bressani, 1997). The improvement of the protein's quality is achieved through the supplementation method, which consists of a mixture of proteins to increase its biological effect. The biological availability of the amino acids is the quantity of a substance that reaches the tissue cells, making it capable of modifying the metabolism of those cells.

TABLE 7.2 Values (Wet Samples) of Proteins, Lipids, Fibers, and Carbohydrates in Citric Waste and in a Control Animal Food

Parameters ($X \pm \delta$)	Lemon Waste	Wet Peel	Dust Peel	Witness Animal Food
%Proteins	5.65 ± 0.28	7.14 ± 0.22	8.57 ± 0.39	7.67 ± 0.28
%Total lipids	0.53 ± 0.01	1.29 ± 0.28	1.29 ± 0.04	1.76 ± 0.05
%Fibers	1.16 ± 0.01	25.33 ± 0.34	30.93 ± 1.10	5.19 ± 0.01
%Carbohydrates	16.02 ± 0.03	4.42 ± 0.17	59.09 ± 1.08	49.92 ± 1.06

Note: X, average values of the experiments realized; δ, standard deviation.

Even though the content of proteins in the lemon waste is low, up to 26%, than the one in the control animal food, it can be considered that these proteins of vegetal origin combined with those of animal origin still provide all the essential amino acids. In the samples of peel dust and wet peel, the content of proteins is similar to the control value, though it would be necessary to make supplementation with proteins of animal origin due to the supply of essential amino acids from the latter that are not found in vegetal origin.

The amount of alimentary fiber is quite significant in the sample of dust peel, coincident with the total lipids. These last components would be easily facilitated due to their low economic cost.

7.4 CONCLUSION

Up to the present moment, the main market of the dried lemon peel was the exportation for the obtention of pectins. Currently, this market is restricted as it is being stocked by other products, so it is possible that in the future the peel is not dried and the wet peel is obtained as a subproduct. Thus, this waste is taken a raw material for this thesis. The results confirmed that the industrial wastes analyzed could be used for animal balanced food, making supplementation of the nutrients in shortage. The use of these supplementary forages to the pastures could lower environmental problems using organic waste as productive resources in the industry, which would increase the socioeconomic efficiency in the NOA area.

KEYWORDS

- **citrus industry**
- **byproducts**
- **nutritional study**
- **industrial waste**
- **balanced feed for animals**

REFERENCES

Bressani R. Protein supplementation and complementation. In: *Evaluation of Proteins for Humans.* Bodwell C. E. (Ed.). AVI Publishing Co.: Connecticut, 1977; p. 240.

Cisint J.C.; Martin G.O.; Toll Vera J.R. Valor nutricional de subproductos de la industria citricola de Tucuman y suposible utilización en alimentación animal. *Revista Argentina de Producción Animal*, 2007.

Coppo J.A.; Mussart de Coppo N.B. Bagazo de citrus como suplemento invernal en vacas de descarte.—46 Suplementación en general en Rumiantes. Sitio Argentino de Producción Animal, 2006. [Online]. http://www.produccion-animal.com.ar/informacion_tecnica/suplementacion/78-bagazo_citrus.pdf (accessed August 7, 2011).

Garcia Garrido J.; Rodriguez Lopez A.D. *Industrias Químicas y Agroalimentarias: Análisis y Ensayos.* Alfaomega: México, 2004.

Latimer G. W. *Official Methods of Analysis of AOAC International*, 19th ed. AOAC International: Maryland, 2012.

Lezcano N. Dieta citrusvil, 2010.

CHAPTER 8

MICROENCAPSULATION AS A TECHNOLOGICAL ALTERNATIVE IN THE FOOD INDUSTRY FOR CONSERVATION OF BETALAINES

JUAN ANTONIO UGALDE-MEDELLÍN[1], LLUVIA ITZEL LÓPEZ-LÓPEZ[2*], JUAN GUZMÁN-CEFERINO[3], SONIA YESENIA SILVA-BELMARES[1], CRISTÓBAL NOÉ AGUILAR[1], and JANETH MARGARITA VENTURA-SOBREVILLA[4]

[1]*School of Chemistry, Autonomous University of Coahuila. Saltillo, Coahuila 25280, México*

[2]*Institute of Research in Desert Areas, Autonomous University of San Luis Potosi, San Luis Potosi 78377, Mexico*

[3]*Agricultural Sciences Academic Division, Juárez Autonomous University of Tabasco, Tabasco, Mexico*

[4]*School of Nutrition, Autonomous University of Coahuila Piedras Negras, Coahuila 26090, México*

[*]*Corresponding author. E-mail: lluvia.lopez@uaslp.mx*

ABSTRACT

Nowadays, it is widely recognized that consuming fruits and vegetables brings huge benefits to human health due to their content of different biomolecules (phytochemicals).

Betalains are water-soluble biomolecules derived from the condensation of betalamic acid with a primary or secondary amine. Due to differences in its structural configuration, betalains are divided into two groups: betacyanins and betaxanthins; both groups have optical properties as they possess two chiral centers at C-2 and C-15 with conjugated double bonds and present the maximum light absorption at 480 nm for betaxanthins and 536 nm for betacyanins. These compounds also possess colors; this feature has led to their wide use in the food area as natural dyes and even in the cosmetics industry. The synthetic dyes have been associated with a number of diseases and disorders; thus, the need for natural products to avoid adverse effects has arises. Betalains belong to a group of five natural coloring additives commonly used in the food industry, especially in meats, dairy products, dehydrated drinks, cold drinks, and jellies. The potential of betalains as food additives is enhanced when they are microencapsulated because this can prevent their degradation and maximize their storage life.

8.1 GENERAL ASPECTS

In nature, it is possible to find a great variety of molecules within plants, which among their many functions we find act as pigments and protectors of plant health; however, these properties can be exploited to maximize their potential; among these phytomomolecules, polyphenols, flavonoids, carotenoids, anthocyanins, etc., are included, which have both antioxidant activity and the ability to prevent some types of cancer (Hurtado and Pérez, 2014). Owing to this, the phytomolecules have attracted the attention of industries such as pharmaceuticals, cosmetics, and food industries (Nava, 2010). Due to the advances in organic synthesis, the scientists have tried to replicate this type of molecules for different applications, such as the case of synthetic dyes, which in the 1950s took great popularity; nevertheless, since 1979, more than 700 dyes have been removed from the market by the Food and Drug Administration (FDA) for being harmful to both the environment and human health. In 2008, the use of 36 dyes was recorded in the food and cosmetic industry, seven of which are synthetic dyes (highly controlled) and the rest (26) are from a natural origin (Georgiev et al., 2008); among these natural dyes permitted due to their low toxicity and their capabilities, we found betalains.

In nature, betalains are pigments found in the vacuoles of plant cells; it is known that their presence is abundant in the beetroot (Marañón-Ruíz et al., 2011). Among their features, betalains are hydrophilic molecules, stable at acidic pH (Esquivel, 2004), and powerful natural antioxidants (providing protection against oxidative stress and inflammatory processes).

Their presence in the plant kingdom is limited to plants of the order *Caryophyllales*, with colors like yellow, red, and violet; other types of natural dyes often used in the food industry are anthocyanins, belonging to the *Molluginaceae* order with a colors ranging from red and blue to violet. Betalains and anthocyanins are mutually exclusive pigments, and so never found together in the same plant; this is because the enzymes responsible for the production of betalains are not present in the plants that produce anthocyanins (Gandía-Herrero and García-Carmona, 2013).

8.2 NATURAL PIGMENTS

Currently, trends such as the use of molecules from natural sources, like as fruits, vegetables, oilseeds, and herbs, have been the focus of many studies for the development of functional foods (Lee et al., 2002); in the food industry, the use of synthetic pigments has decreased due to toxicity problems (Puértolas et al., 2013), intolerance reactions, and allergies (Soriano-Santos et al., 2007) has been associated with carcinogenic processes (Boo et al., 2012). The use of natural dyes has presented a great alternative to this problem because it is possible to take advantage of its nutritional and functional capabilities as antioxidant, antiproliferative, and radio-protective effects, and even some authors recommend them for the treatment of diabetes and prediabetes (Chandrasekhar et al., 2015; Lee et al., 2002). So, it has greatly increased the demand for foods with better nutritional qualities (Bello-Gutiérrez, 2012).

The most common plant pigments are carotenoids, chlorophylls, and anthocyanins (Stintzing and Carle, 2004). Nevertheless, betalains are among the natural pigments of major interest in the food industry and are recognized as nutraceutical molecules potentially beneficial to human health. The term "nutraceuticals" was coined by Stephen De Felice (founder of the Foundation for Innovation in Medicine, USA, 1989) (Pérez-Leonard, 2006) who defined nutraceuticals as "any food, or parts of a food, which provides health benefits, including the prevention and treatment of disease" (Scarafoni et al., 2007).

This term may cause some confusion due to its similarity with the definition of a "functional food", which, in turn, has a multitude of definitions (Doyon and Labrecque, 2008) and can be summarized as follows: a functional food is one with some components or nutrients, which has an effect on the consumer; it is also possible to define it as a food that in the physiological or psychological form generates an effect on the consumer beyond the traditional nutritional effect (Taper and Roberfroid, 1999). Therefore, it is possible to see that the term "nutraceutical" only encompasses natural molecules with pharmacological power, applied inside or outside a food system (clearly not excluded the possibility). Despite being molecules widely used in the food industry as colorants, it was only in 1957 that Wyler and Dreiding first isolated crystals with a red–violet hue (which later became to known as betacyanins or betanins) obtained from the root of *Beta vulgaris* or table beet, from which they were given the name betalains since beet has been the most widely used commercial source (Sáenz and Berger, 2006).

Synthetic dyes have been associated with a number of diseases and disorders (such as attention deficit hyperactivity disorder). Thus, there lies the need for finding natural products devoid of adverse effects (Robert de Mello et al., 2015). The annual production of the main natural sources of betalains is as follows: first, beetroot with 20–70 t/ha per year with a betalain content of 0.4–20 mg/g; second, cactus pear (fruit) with an annual production of 45 t/ha per year with a betalain content of 0.8 mg/g; third, swiss chard (petiole) with an annual production of 35–40 t/ha per year with a relatively low betalain content of 0.04–0.08 mg/g; and finally, pitaya fruit with an annual production of 7–19 t/ha per year with a good betalain content of 0.32–0.41 mg/g (Khan and Giridhar, 2015).

The main source of betalains is the red beet; this tubercle has been established in the market as the source of the oldest and most abundant red food colorant, called betanin, which is known as E-162 by the Code of Federal Regulations in the European Union and the FDA in the United States; this pigment can be found in chilled dairy products, like fruit yogurts, juices, confectionery items, ice creams, syrups, sausages, and processed meats (Aberoumand, 2011; Kaimainen et al., 2015). Nevertheless, this pigment (betanin) has a typical earthy flavor caused by the presence of geosmin and high nitrate concentrations, both of which are associated with the carcinogenic process; furthermore, there is a risk that the raw material in red beet carries earth-bound microorganisms (Aberoumand, 2011). For that reason, there has been an increasing demand for other alternative betalains sources. Fortunately, in Mexico, there are many potential sources, like the cactus fruit belonging to the *Caryophyllales* order, such as *Aizoaceae, Amaranthaceae, Basellaceae, Cactaceae, Chenopodiaceae, Didieraceae, Holophytaceae, Nyctaginaceae, Phytolaceae,* and *Portulacaceae* besides the species such as *Amaranthus, Hylocereus* (Vaillant et al., 2005), *Opuntia* (Khatabi et al., 2013), and *Myrtillocatus* (Guzmán-Maldonado et al., 2010), which present high presence of biomolecules with biological activity (García-Cruz et al., 2012; Seigler, 2012) and several technological and sensorial advantages in comparison to red beet (Otálora et al., 2015). Betalains have been isolated from prickly pear, pokeweed, and bougainvilleas, as well as some fungi such as *Hygrocybe* (Khatabi et al., 2013). Recent studies have identified alternatives sources of betalains such as the pitaya shell, which is considered a waste of consumption is usually discarded but rich in betalains (Robert de Mello et al., 2015).

In Mexico, several plant species produce betalains pigments, among these are *Myrtillocactus geometrizans* or Garambullo, which is a Mexican cactus distributed mainly in the states of Hidalgo, San Luis Potosi, Querétaro, and Guanajuato, being these also its main producers (Topete-Viniegra, 2006); this is a Cactaceae easily grown in arid and semiarid zones on mountainsides and lowlands (Reynoso et al., 1997). Some cacti such as *Mammillaria candida Scheidweiler*, which is a Mexican cactus, has been found to have a high generation of pigmented calli in tissue culture in vitro (Santos-Díaz et al., 2005). May pitaya (*Stenocereus griseus* H.) is a columnar cactus typical of Mexico. Its fruits are globose or ovoid berries, with deciduous thorns; the pulp can be orange, red, or purple (García-Cruz et al., 2012). Mexico is one of the centers of origin and dispersion of the genus *Opuntia*, whose fruit is typically known as prickly pear and contains carotenoids and betalains, providing color from pale yellow to orange, being attractive for the preparation of various products (Escuela Nacional de Agricultura (Mexico) et al., 2010). Pitaya (*Hylocereus undatus*) is a native fruit of Mexico and Central and South America. Pitaya peel is considered a residue from the consumption and processing of the fruit usually discarded. However, this residue can be used as a raw material for the extraction of pigments due to the presence of betalains (Robert de Mello et al., 2015). The "jiotilla" is the fruit of the arborescent cactus *Escontria chiotilla*, grows wild in arid zones of the Mexican territory within the Mixteca Baja Oaxaqueña, Guerrero, Michoacán, and south of Puebla, and produces edible red fruit pulp with a bittersweet flavor (Soriano-Santos et al., 2007). Regardless of their source of production, betalains have important pigmentation properties that in some cases are even more potent than artificial colorants, with the advantage over other natural pigments such as anthocyanins to be stable at acid pH in a wide range of 3–7 (Sánchez-González, 2006); for that reason, it is of great interest for the food industry to find other sources of exploitation of these kinds of pigments (Soriano-Santos et al., 2007).

Currently, more than 75 different betalains have been identified, of which around 25 belong in the subgroup of betaxanthins and the other 50 are part of the betacyanins; an important point to note is that both species of betalains (betacyanins and betaxanthins) are present in each plant species, in which the different spectrum of color is determined by the betacyanins and betaxanthins proportions. Some of the isolated and more studied betalains with red–violet color are betanin, isobetanin, bethanidine, and isobetanidina and some of the yellow pigments known are vulgaxanthin I and II (Nava, 2010; Serrano et al., 2012). These kinds of molecules have some characteristics that we need to focus on at the moment to work because they are biodegradable by light, oxygen, and heat and are stable at the acidic pH range of 3–6; for that reason, their main applications in the food industry are in pastries, drinks, sauces, jellies, and meat products (Chethana et al., 2007; Olea-Serrano et al., 2012).

8.3 BETALAIN'S COLOR

The color of the food itself is important, as it may affect future choices and purchase decisions. Colors may also significantly affect the perceptions of foods and form expectations of how they would smell or taste (Kaimainen et al., 2015). Food ingredients are the focus of public interest; it is becoming increasingly important to meet consumer's expectations for natural, healthy products, and hence, the search for new plant-derived colorants for the food industry is still necessary. However, the higher stability of synthetic colorants compared to that of natural alternatives is a challenge that must be overcome. To meet the growing demand for natural colorants, new pigment sources are being sought, practically betalains have been used to color foods such as yogurt, confectionery, ice creams, syrups, sausages, and processed meats.

Betalains absorb radiation in the visible range between 476 and 600 nm (Al-Alwani et al., 2015); their color is directly related to the presence of certain organic functional groups (Gutiérrez, 2000). Betalains are divided into two main groups ,betacyanins having a range of colors from red to violet with a $\lambda_{max} \approx 540$ nm and the betaxanthins having a range of yellow colors with a $\lambda_{max} \approx 480$ nm (Al-Alwani et al., 2015). In 1980, Schwartz and Von Elbe scientists determined the molar absorbance of betanins (the betacyanin from beet) with an ideal molar extinction coefficient; they obtained a value of 1120% /mol cm, which has been employed to calculate the real concentration of betanin extracts (and any other betacyanin) (Soriano-Santos et al., 2007). The chromophore of betalains can be described as a protonated compound 1,2,4,7,7-penta and 1,7-diazaheptametin system. This system is also involved in the antioxidant activity of betalains (Marañón-Ruíz et al., 2011).

The concentration of betacyanins and betaxanthins in an extract can be determined thanks to Nilson's equation (Eq. (8.1)), who in 1970 determined spectrophotometrically the concentration of total betalains, as well as betaciananias and betaxanthins as a ratio of the absorbance (A) of the sample at 483 and 538 nm for betaxanthins and betacyanins, respectively, their respective molecular weights (MW = 550.5 g/mol betacyanins and MW = 339.3 g/mol for betaxanthins), their molar extinction coefficient (60,000 L/mol×cm for betacyanins and 48,000 L/mol×cm for betaxanthins), their ideal extraction coefficient (E = 1120%/mol cm for betacyanins and E = 750%/mol cm for betaxanthins), their respective dilution factor (FD), the length of the trajectory of the cell (L = 1 cm), and the volume (V) used (García-Cruz et al., 2012; Hernández, 2008; Nilsson, 1970; Soriano-Santos et al., 2007).

$$\frac{\text{mg pigment}}{100\text{ g}} = \frac{(A \times \text{FD} \times \text{PM} \times V)}{E \times L} \tag{8.1}$$

8.4 BIOLOGICAL ACTIVITY

The ability of the biomolecules to possess a favorable biological activity on the human body is not an entirely new fact; for example, the Mexican indigenous communities have been using the leaves and fruits of *Opuntia* (prickly pear) for its medicinal benefits, which are useful in the treatment of atherosclerosis, diabetes, gastritis, and hyperglycemia (Lee et al., 2002). Among its beneficial properties, the highlighted is the antioxidant activity that protects it against the damage by free radicals or ROS (radical oxides) such as superoxide anion, hydroxyl radicals, and H_2O_2, which cause a decrease in the nutritional food value and affect its appearance and safety (Pangestuti and Kim, 2011). These free radicals attack macromolecules such as membrane lipids, proteins, and DNA, leading to many health problems such as cancer, diabetes mellitus, aging, and neurodegenerative diseases. This topic is very important for food preservation because it seeks to prevent or reduce lipid oxidation, which causes rancidity thereof in addition to the formation of undesirable secondary products of lipid peroxidation (Kayodé et al., 2012; Sánchez-González, 2006). There exists a direct relationship between the chemical structure and antioxidant activity of the betalains, which gives them properties for removing free radicals; this usually increases with the number of hydroxyl and amine groups and depends on the position of these groups as well as the presence of glycosylated groups. The presence of hydroxyl groups at the C-5 position of the molecules of betalains improve the antioxidant activity, and more glycosylations reduce the biological activity (Cai et al., 2005); for example, it was demonstrated that the betalains are capable of inhibiting the peroxidation of linoleic acid and the oxidation of LDL, inhibiting, even at lower concentrations in the diet, the skin and liver tumor formation in mice, and protecting against the effects of gamma radiation. In addition, plasma concentrations of betalains after ingestion, even at low concentrations, are sufficient to promote their incorporation into the plasma, LDL, and erythrocytes, which are then protected from the oxidative damage and hemolysis(Gandía-Herrero and García-Carmona, 2013; Gandía-Herrero et al., 2010). Besides that the betalains can be regarded as cancer preventive agents, they also exhibit good antiviral and antimicrobial activities; moreover, betaxanthins can be used as a rich source of essential amino acids in the diet (Loginova et al., 2011).

8.5 BETALAINS APPLICATIONS IN THE FOOD INDUSTRY

Betalains have been widely used as additives dyes of natural origin, mainly in products consisting of soy, meats, desserts, dairy products, etc. Betalains can be used to extend the shelf life of other foods because of their ability to trap free radicals and inhibit lipid peroxidation. Also, the consumption of products containing betalains provides protection against certain oxidative stress-related diseases (Topete-Viniegra, 2006). As mentioned before, there are different factors that affect the stability and half-life (Garibay et al., 1993; Loginova et al., 2011; Topete-Viniegra, 2006; Ugaz, 1997). However, betalains have been successfully used to color ice creams (Kumar et al., 2015). Also, betalains from *Opuntia stricta* have been used to color yogurt and soft drinks with good color stability after one month of storage under refrigerated conditions (Obón et al., 2009). Betacyanins from *Amaranthus* were used to color jellies, ice creams, and beverages with good color stability at low temperatures after 12 or 18 weeks, but with inferior stability at room temperature (Cai and Corke, 1999).

However, despite the studies on its antioxidant activity, importance, and perspectives, there is a little information about the pharmacokinetics of betalains in humans; however, it is known that the urinary excretion of betalains is less than 0.30% of the administered dose by orally, indicating either the low bioavailability of these pigments or there could be a different route of elimination. In recent investigations with different food matrixes, it was found that the commercial pigment has a low half-life in plasma, whereas in cactus fruit consumption, the presences of this compound in plasma was more than 8 h. Thus, the food matrix plays a role in the stability, bioaccessibility, and bioavailability of betalains; for that reason, there have been different investigations on the addition of betalains to different food matrixes, such as milk, ice cream, cookies, and gummy confections (Celli and Brooks, 2016).

However, betalains are not well studied in comparison to other natural pigments like carotenoids and anthocyanins and information regarding the effects of processing conditions on their physicochemical properties is scarce. Nevertheless, betalains have some advantages over these pigments, thanks to their regeneration capability and stablility at a wider pH range (between pH 3 and 7) compared to anthocyanins (Celli and Brooks, 2016). For example, the yogurt drink had a very attractive red-purple hue for consumers, whose did not change after a month of cooling, but this stability would be higher if applied via microencapsulation technology (Vergara Hinostroza, 2013). Betalains show a linear relationship between its degradation and the presence of different factors at a given time, which leads to first-order degradation kinetics (very similar to the relationship of compounds such as anthocyanins; Hurtado and Pérez, 2014; Sagdic et al., 2013). A study on some extracts from Tuna with glucose syrup shows stability for a month at room temperature (Obón et al., 2009); this behavior even occurs in the case of juices containing betalains (Reynoso et al., 1997; Rosario-Castellar et al., 2003). Microencapsulation is often used to increase the stability and bioefficacy of some natural compounds as is the case of betalains. The studies suggest that microencapsulation of betalains with gum acacia, maltodextrin, inulin, storing under low aqueous activity conditions, less than 0.5, 0.7, and 0.8 has restored pigment content up to 45 days (Manchali et al., 2013). Extracts of *Opuntia lasiacantha* with maltodextrin as an encapsulant stored for 24 weeks showed a pigment retention of 89.2 ± 3.9% (Díaz-Sánchez et al., 2006). The stability of betalains depends not only on the type of their natural source but also on the food matrix in which they are being used. During the processing of food, some steps in the process may have negative effects on the stability of pigments; fortunately, the food matrix and other ingredients used in food may have positive effects on the stability of the betalains. For example, the addition of antioxidants like ascorbic acid and isoascorbic acid to food formulations enhance the stability of pigment tanks by the removal of oxygen; the addition of chelating agents such as citric acid and EDTA is also known to increase the stability of pigments, and enzymes like glucose oxidase and β-cyclodextrin are known to provide the best stability to pigments through absorption of water and removal of dissolved oxygen, respectively (Manchali et al., 2013). The stability of betalains is strongly influenced by the pH, temperature, light, water activity (aw), and the presence of oxygen, a factor restricting their use as a colorant in foods. According to these characteristics, betalains can be used in foods with a short shelf life, processed with a minimal heat treatment, and packed and placed in a dry place under low light, oxygen level, and moisture conditions (Sánchez-González, 2006). The degradation of betalain molecules is accompanied by a change in its color. Within this context, the heating treatments cause degradation of betalains by isomerization, decarboxylation, or cleavage, resulting in a gradual reduction of red or violet and the appearance of a light brown color. Cleavage of betanin and isobetanina can also be induced by bases, generating molecules as the bright yellow betalamic acid and the colorless cyclo-dopa-5-*O*-glycoside (CDG) (Mello et al., 2015). Under mild alkaline conditions, betanin is degraded to betalamic acid and CDG. These two products are also formed during the heating of acidic solutions of betanin or during the thermal processing of products containing betanin. The degradation of betanin to betalamic acid and CDG is reversible, and therefore, after heating, a partial regeneration of the pigment occurs. The proposed feedback mechanism involves the condensation Schiff base of the aldehyde group of betalamic acid and nucleophilic amine from CDG; betanin regeneration is maximized at an intermediate pH range (4.0–5.0). Betacyanins, due to a chiral center at C-15, have two epimers. The epimerization occurs by acid or heat. It is hoped, therefore, that during heating of foods containing betanin, the betanin relationship with isobetanina also increases. It has also been observed that when heating occurs in aqueous solutions, the decarboxylation of betanin occurs. The addition of ascorbic acid, isoascorbic acid, and citric acid increases their stability to heat, at pH 4 and 6. Moreover, the hydrolysis catalyzed by enzymes produces only bethanidine (Castro-Muñoz, 2006; Sánchez-González, 2006).

In food, the stability of phytocompounds such as betalains is strongly influenced by the chemical composition of the species of origin, the factor that accelerates the process of decarboxylation of betalains (Moreno et al., 2007). Different strategies have been proposed to prevent or reduce the degradation of betalains, such as the use of ascorbic acid, citric acid, tetracthyl orthosilicate, and alternative solvent systems and encapsulation techniques such as spray drying (Celli and Brooks, 2016). Encapsulation promotes easy handling, offers better solubility and stability, improves flow properties, reduces dusting when the nutrients are added to dry mixtures, prevents lumping, and enhances the shelf life of the encapsulated compounds. In food systems, the microcapsule protects the core material from degradation by reducing its reactivity to environmental conditions; thus, the stabilization of betalain pigments may boost its use as natural bioactive and coloring molecules in the food industry and promote their application in other areas (Ravichandran et al., 2014).

8.6 MICROENCAPSULATION OF BETALAINS

As already mentioned, there are multiple benefits obtained by consuming natural products, but the problem arises when the incorporation of these phytomolecules to food, for example, juice, reduces its nutritional value due to different environmental degradation conditions such as the change of pH, oxygen, temperature, light, etc., (Cardoso-Ugarte et al., 2014). Betalains show a linear relationship between their degradation and the presence of different factors at a given time, which leads to the first-order degradation kinetics (very similar to the relationship of compounds such as anthocyanins) (Hurtado and Pérez, 2014; Sagdic et al., 2013). So, it is necessary to find solutions to these problems. Currently, an alternative is the technology of microencapsulation (Parize et al., 2010).

Microencapsulation process wherein the solid particles result in micrometric size, the size of the microparticles range between 15 and 100 μm; it is important to note that microcapsules larger than 100 μm are detectable in the mouth and those below 15 μm do not provide sufficient protection (Pérez-Leonard et al., 2013). The microencapsulation technique was developed between 1930 and 1940 by the National Cash Register (Popplewell et al., 1995); however, until recent years, this technique has become popular for its implementation in different areas such as pharmaceuticals, cosmetics, and food industries (Fang and Bhandari, 2010). Bioactive compounds in the liquid, solid, or gaseous state are coated with a porous polymer film that creates a network with hydrophobic and/or hydrophilic properties (Fuchs et al., 2006), thus protecting the bioactive components from the activity of oxygen, water, light, among other conditions to improve their stability (Saénz et al., 2009). Depending on the process of the core and wall materials, different morphologies can be obtained; the main ones being mononuclear capsules, which have a single nucleus surrounded by a layer, and aggregates, where the capsules have many nuclei. For both cases, the particles may be spherical or irregular (Esquivel-González et al., 2015). The coated material is called the active material or core, and the coating material is called the shell wall or encapsulating vehicle (Madene et al., 2006). The protective material, being usually a semipermeable polymeric membrane, must have certain chemical properties that determine the behavior of the capsule, as well as the process of release of the material, its application, and storage conditions. The ideal characteristics of a coating are low viscosity at high concentrations, low hygroscopicity to facilitate handling and avoid agglomeration, the ability to emulsify, unreactive with the core material. To ensure the effectiveness of the process, it must be taken into account whether the coating is soluble in common food solvents or the final food product (Pérez-Leonard et al., 2013; Pérez-Alonso et al., 2009). The selection of the encapsulating material is very important in the process of spray drying since it intervenes directly with the stability of the product, the degree of protection of the active core, and the efficiency of the process. Among the main agents used as coatings in the process of spray drying are gums; sugars; natural polysaccharides such as hydrophobic starches; maltodextrin corn syrup; pectin, carboxymethylcellulose; Arabic gum; guar gum; chitosan; sodium alginate; lipids; proteins such as gelatin, casein, soy, sodium caseinate, whey protein, and mixtures thereof (Parize et al., 2010; Ravichandran et al., 2014); and synthetic polymers (Martínez, 2015) and proteins, among which caseinate by its high emulsifying capacity, high colloidal stability, stability to flocculation and/or coalescence allows obtaining capsules of sizes as small as 2 μm or less (Guerrero-Haberl et al., 2011). The presence of calcium ions favors the intermolecular interaction of the caseins due to the presence of phosphoserine groups in the proteins, which produce strong ionic bonds (Casanova and Cardona, 2004). The main techniques used for microencapsulation can be spray drying, freeze drying, air suspension coating, extrusion, spray cooling and spray chilling, centrifugal extrusion, rotational suspension, and simple and complex coacervation; nevertheless, still the most used and studied technique is the spray drying process. The spray drying process involves three stages: preparation of the emulsion, homogenization, and atomization, so this process has been used for decades to encapsulate food ingredients (Parize et al., 2010; Ravichandran et al., 2014). With this technique, the betalains microencapsulated express a pseudo-first-order degradation, showing better stability in a long period of storage (Saénz et al., 2009). The spray drying process has been used to improve the taste and stability of drugs and acts as a barrier against odors and flavors. In the case of drugs, a release takes place in the stomach or intestine, allowing maximum absorption of the compounds with minimal adverse reactions. The encapsulation method selection will depend on the particle size required, physical properties of the encapsulant, the substance to be encapsulated, the proposed applications of the encapsulated material, desired release mechanism, and cost (Martín-Villena et al., 2009). It has recently been found that natural extracts have greater power once antioxidant microencapsulated, thanks to the interaction between the cover and microencapsulated betalain (Salazar-González et al., 2009). During and after drying, the physical properties of microcapsules change because of the loss of their moisture content. For that reason, the major research studies focus on the characterization of the physical properties as the changes in volume, area, shape, changes in the surface are directly related to the structure of the drying material (Pérez-Alonso et al., 2009). The quality of the microcapsules produced by spray drying depends directly on the physicochemical characteristics of the solution, such as viscosity, flow velocity, etc., the working conditions in the equipment, such as the temperature, pressure, and airflow, the contact between hot air and the droplets in the drying chamber depending on the type of flow, either in parallel currents or countercurrents, and no less important the type of the atomizer used. The greater the energy supplied, the finer the drops formed and therefore the smaller the microparticles obtained. Experimentally, it has been proven that by using the same amount of energy, the particle size increases when the feeding speed increases. In the same way, the particle size increases when the food liquid presents high surface tension and viscosity. Among the most important advantages of this technique are the use of short production times, economic feasibility, and the use of low temperatures (Esquivel-González et al., 2015).

Most studies on encapsulation have focused on analyzing the conditions of the drying process as well as the morphological characteristics of the microcapsules obtained; however, a few studies have focused on the different oxidative processes that occur within the microcapsules and that lead to the rancidity of the encapsulated product (Carrillo-Navas et al., 2010; Fang and Bhandari, 2012). In this type of processes, it is necessary to control the water activity and the moisture content on the polymeric cover (Carrillo-Navas et al., 2010). For this reason, it is necessary to consider the glass transition temperature (T_g) as a critical parameter for the different undesirable processes that could occur during the heating of the material, such as collapse, stickiness, caking, agglomeration, and recrystallization phenomena. In this sense, a change from a glassy to a rubbery state can occur as a consequence of an increase in the temperature or the water content during its storage (Guadarrama-Ledezma et al., 2014).

Microencapsulated colorants can be used commercially as a food dye, mainly for children products, like yogurt, ice creams, sweets, glace cherries, soft drinks, and jam syrups (Özkan and Bilek, 2014). Betalain's microencapsulation has been little studied, but there are some studies on spray drying with different natural sources. However, most of the studies focus on beetroot extracts. The encapsulated betalains present a high interaction with most encapsulating polymers; this is due to their cationic characteristics producing strong electrostatic interactions. In addition to this, hydrogen bonds are produced that allow a high cohesion between the internal occluders of the particular one. In addition, the presence of pulp components such as mucilage (in crude extracts) plays an important role in the microencapsulation process as they protect betalains from different degradative processes inherent to the heating process; however, some sugars may play a harmful role in the storage process for long periods. It was previously mentioned that water activity plays a crucial role in the quality of microcapsules, but this factor is very important in the stability of betalains since the pigment becomes more unstable as the water activity and moisture content of the food increase. Therefore, a decrease in water activity corresponds to lower degradation of betanin, as a result of the decrease of different internal hydrolytic reactions of the capsule. Values lower than 0.21 favor the stability of microcapsules (Esquivel-González et al., 2015). The microencapsulation process has been used to increase their stability and bioefficacy; for example, betalain pigments microencapsulated with gum acacia are stable under low moisture conditions for more than 45 days (Pitalua et al., 2010). Betalain pigments obtained from cactus pear juice by ethanolic extract microencapsulated with maltodextrin and inulin are stable at 60 °C for up to 44 days (Saénz et al., 2009); in betalains obtained from beetroot and encapsulated with maltodextrins by the spray drying method, it was observed that increasing the air inlet temperature decreases the yellow pigment (Janiszewska and Wáodarczyk, 2013). Betacyanins' drying yield, color yield, and color strength are stable in outlet temperatures of 160 °C using glucose syrup (Obón et al., 2009); there are some studies on applying betalains to food matrixes like yogurt; the addition of microparticles of bioactive compounds obtained from *Opuntia ficus-indica* or *O. stricta* maintains a high retention rate and betalains as a deep purple color throughout the study period (30 days at 5 °C) (Obón et al., 2009; Torres, 2008). Betalains microencapsulated with guar gum by the freeze-drying technique have demonstrated higher stability compared to those by the spray-dried technique (Ravichandran et al., 2014). Despite the few works on betalain's microencapsulation, the results of these studies suggest that encapsulation is a promising method for the stability of betalains, increasing their storage life, and the possibilities of their application in the food industry (Manchali et al., 2013).

8.7 FINAL COMMENTS

Betalains are versatile molecules with applications in different areas from food to pharmacological industry, but their main application is as food additives for their staining capabilities and antioxidant properties. However, their lability has been a problem that has diminished their application in recent years; for that reason, the implementation of technologies that prevent their degradation like the microencapsulation process is one of the most promising methods for the food industry because the color and the bioactivity are preserved and more important it is a cheap and efficient technique. In conclusion, the contributions to mankind by the plants and their biomolecules are enormous and cannot be denied; however, it is necessary to study and find new compounds capable of overcoming current problems as other molecules did in the past.

KEYWORDS

- **microencapsulation**
- **conservation**
- **betalains**

REFERENCES

Aberoumand, A. (2011). A review article on edible pigments properties and sources as natural biocolorants in foodstuff and food industry. *World Journal of Dairy & Food Sciences*, *6*(1), 71–78. Retrieved from https://pdfs.semanticscholar.org/ba8b/d411277afa8e0baf9c875bbe42e93a20a922.pdf

Al-Alwani, M. A. M., Mohamad, A. B., Kadhum, A. A. H., & Ludin, N. A. (2015). Effect of solvents on the extraction of natural pigments and adsorption onto TiO_2 for dye-sensitized solar cell applications. *Spectrochimica Acta Part A: Molecular and Biomolecular Spectroscopy*, *138*, 130–137. https://doi.org/10.1016/j.saa.2014.11.018

Boo, H. O., Hwang, S.-J., Bae, C.-S., Park, S.-H., Heo, B.G., & Gorinstein, S. (2012). *Extraction and characterization of some natural plant pigments*. In *Industrial Crops and Products* (Vol. 40). Elsevier B.V. https://doi.org/10.1016/j.indcrop.2012.02.042

Cai, Y., & Corke, H. (1999). Amaranthus betacyanin pigments applied in model food systems. *Journal of Food Science*, *64*(5), 869–873. https://doi.org/10.1111/j.1365-2621.1999.tb15930.x

Cai, Y. Z., Sun, M., & Corke, H. (2005). Characterization and application of betalain pigments from plants of the *Amaranthaceae*. *Trends in Food Science and Technology*, *16*(9), 370–376. https://doi.org/10.1016/j.tifs.2005.03.020

Cardoso-Ugarte, G. A., Sosa-Morales, M. E., Ballard, T., Liceaga, A., & Mart, M. F. S. (2014). Microwave-assisted extraction of betalains from red beet (*Beta vulgaris*). *Food Science and Technology*, *59*, 276–282. https://doi.org/10.1016/j.lwt.2014.05.025

Carrillo-Navas, J. H., Cruez-Olivares, J., Barrera-Pichardo, J. F., & Pérez-Alonso, C. (2010). Estabilidad térmica oxidativa de microcápsulas de saborizante de nuez. *Superficies y Vacío*, *23*, 21–26. Retrieved from http://www.redalyc.org/articulo.oa?id=94248264005

Casanova, Y., Herley, F., Cardona, T., Sara, C. Emulsiones O/W estabilizadas con caseinato de sodio: efectos de los iones calcio, concentración de proteína y temperatura. VITAE, *Revista de la Facultad de Química Farmacéutica*, *11*, 13–19.

Castellar, R., Obón, J.M., Alacid, M., Fernández-López, J.A. (2003). (2003). Color properties and stability of betacyanins from Opuntia fruits. *Journal of Agricultural and Food Chemistry*, *51*(9), 2772–2776.

Castro Muñoz, R. (2014). *Efecto del secado por aspersión en la estabilidad de componentes bioactivos de tuna morada (Opuntia ficus-indica)*, Master Dissertation. Instituto Politécnico Nacional, México.

Celli, G. B., & Brooks, M. S.-L. (2016). Impact of extraction and processing conditions on betalains and comparison of properties with anthocyanins—a current review. *Food Research International*. https://doi.org/10.1016/j.foodres.2016.08.034

Chandrasekhar, J., Sonika, G., Madhusudhan, M. C., & Raghavarao, K. S. M. S. (2015). Differential partitioning of betacyanins and betaxanthins employing aqueous two phase extraction. *Journal of Food Engineering*, *144*, 156–163. https://doi.org/10.1016/j.jfoodeng.2014.07.018

Chethana, S., Nayak, C. A., & Raghavarao, K. S. M. S. (2007). Aqueous two phase extraction for purification and concentration of betalains. *Journal of Food Engineering*, *81*(4), 679–687. https://doi.org/10.1016/j.jfoodeng.2006.12.021

Díaz Sánchez, F., Santos López, E. M., Kerstupp, S. F., Ibarra, R. V., & Scheinvar, L. (2006). Colorant Extraction from red prickly pear (*Opuntia lasiacantha*) for food application. *Journal of Environmental*, *5 (2)*, 1330–1337. Retrieved from https://pdfs.semanticscholar.org/b1dc/6865432e8331bb348c69c4b2dee5e7e16a15.pdf

Doyon, M., & Labrecque, J. (2008). Functional foods: a conceptual definition. *British Food Journal*, *110*(11), 1133–1149. https://doi.org/10.1108/00070700810918036

Escuela Nacional de Agricultura (Mexico), Colegio de Postgraduados, I., Martínez-Damián, M. T., Rodríguez-Pérez, E., Colinas-León, M. T., Valle-Guadarrama, S., Ramírez-Ramírez, S., & Gallegos-Vázquez, C. (2010). *Agrociencia.Agrociencia* (Vol. 44). Colegio de Postgraduados. Retrieved from http://www.scielo.org.mx/scielo.php?script=sci_arttext&pid=S1405-31952010000700003

Esquivel-González, B. E., Ochoa-Martínez, L. A., & Rutiaga-Quiñones, O. (2015). Microencapsulación mediante secado por aspersión de compuestos bioactivos. *Revista Iberoamericana de Tecnología Postcosecha*, *16*(2), 180–192. Retrieved from http://www.redalyc.org/pdf/813/81343176006.pdf

Esquivel, P. (2004). Los Frutos de las Cactáceas y su Potencial como Materia Prima. *Agronomía Mesoamericana*, *15*(2), 215–219.

Fang, Z., & Bhandari, B. (2010). Encapsulation of polyphenols—a review. *Trends in Food Science and Technology*, 21(10), 510–523.

Fang, Z., & Bhandari, B. (2012). Comparing the efficiency of protein and maltodextrin on spray drying of bayberry juice. *Food Research International*, *48*(2), 478–483. https://doi.org/10.1016/j.foodres.2012.05.025

Fuchs, M., Turchiuli, C., Bohin, M., Cuvelier, M. E., Ordonnaud, C., Peyrat-Maillard, M. N., & Dumoulin, E. (2006). Encapsulation of oil in powder using spray drying and fluidised bed agglomeration. *Journal of Food Engineering*, *75*(1), 27–35. https://doi.org/10.1016/j.jfoodeng.2005.03.047

Gandía-Herrero, F., & García-Carmona, F. (2013). Biosynthesis of betalains: yellow and violet plant pigments. *Trends in Plant Science*, *18*(6), 334–343. https://doi.org/10.1016/j.tplants.2013.01.003

Gandía-Herrero, F., Jiménez-Atiénzar, M., Cabanes, J., García-Carmona, F., & Escribano, J. (2010). Stabilization of the bioactive pigment of *opuntia* fruits through maltodextrin encapsulation. *Journal of Agricultural and Food Chemistry*, *58*(19), 10646–10652. https://doi.org/10.1021/jf101695f

García-Cruz, L., Salinas-Moreno, Y., & Valle-Guadarrama, S. (2012). Betalaínas, compuestos fenólicos y actividad antioxidante en pitaya de mayo (*Stenocereus griseus* H.). *Revista*

Fitotecnia Mexicana, 35(SPE-5), 01–05. Retrieved from http://www.scielo.org.mx/scielo.php?script=sci_arttext&pid=S0187-73802012000500003&lng=es&nrm=iso&tlng=es

García-Garibay, M., Quintero-Ramírez, R., López-Munguía Canales, A. (1993). *Biotecnología alimentaria*. (N. Editores, Ed.). Editorial Limusa, Mexico. Retrieved from https://books.google.com/books?id=2ctdvBnTa18C&pgis=1

Georgiev, V., Ilieva, M., Bley, T., & Pavlov, A. (2008). Betalain production in plant in vitro systems. *Acta Physiologiae Plantarum, 30*(5), 581–593. https://doi.org/10.1007/s11738-008-0170-6

Guadarrama-Ledezma, A. Y., Cruz-Olivares, J., Martínez-Vargas, S. L., Carrillo-Navas, H., Román-Guerrero, A., & Pérez-Alonso, C. (2014). Determination of the minimum integral entropy, water sorption and glass transition temperature to establishing critical storage conditions of beetroot juice microcapsules by spray drying. *Revista Mexicana de Ingeniería Química, 13*(2), 405–416. Retrieved from http://www.redalyc.org/articulo.oa?id=62031508005

Guerrero-Haber, J. R., Ramírez-Perú, A. L., & Puente-Vidal, W. (2011). Caracterización del suero de queso blanco del combinado lácteo santiago. *Tecnología Química, 31(3)*, 313–323.

Bello-Gutiérrez, J. (2000). *Ciencia bromatológica: principios generales de los alimentos*. Ediciones Díaz de Santos, Madrid, España. Retrieved from https://books.google.com/books?id=94BiLLKBJ6UC&pgis=1

Gutiérrez, J. B. (2012). *Calidad de vida, alimentos y salud humana: fundamentos científicos*. Ediciones Díaz de Santos. Retrieved from https://books.google.com/books?id=p6TK3G383pgC&pgis=1

Guzmán-Maldonado, S. H., Herrera-Hernández, G., Hernández-López, D., Reynoso-Camacho, R., Guzmán-Tovar, A., Vaillant, F., & Brat, P. (2010). Physicochemical, nutritional and functional characteristics of two underutilised fruit cactus species (Myrtillocactus) produced in central Mexico. *Food Chemistry, 121*(2), 381–386. https://doi.org/10.1016/j.foodchem.2009.12.039

Hernández, M. G. H. (2008). Efecto del estado de madurez y el almacenamiento sobre la calidad funcional del garambullo (Myrtillocactus Geometrizans). Master dissertation, Universidad Autónoma Metropolitana-Iztapalapa, México.

Hurtado, N. H., & Pérez, M. (2014). Identificación, estabilidad y actividad antioxidante de las antocianinas aisladas de la cáscara del fruto de capulí (Prunus serótina spp capuli (Cav) Mc. Vaug Cav). *Información Tecnológica, 25*(4), 131–140. https://doi.org/10.4067/S0718-07642014000400015

Janiszewska, E., & Wáodarczyk, J. (2013). Influence of spray drying conditions on beetroot pigments retention after microencapsulation process. *Acta Agrophysica, 20*(2), 343–356. Retrieved from http://produkcja.ipan.lublin.pl/uploads/publishing/files/Janiszewska-343-356.pdf

Kaimainen, M., Laaksonen, O., Järvenpää, E., Sandell, M., & Huopalahti, R. (2015). Consumer acceptance and stability of spray dried betanin in model juices. *Food Chemistry, 187*, 398–406. https://doi.org/10.1016/j.foodchem.2015.04.064

Kaimainen, M., Marze, S., Järvenpää, E., Anton, M., & Huopalahti, R. (2015). Encapsulation of betalain into w/o/w double emulsion and release during in vitro intestinal lipid digestion. *LWT—Food Science and Technology, 60*(2), 899–904. https://doi.org/10.1016/j.lwt.2014.10.016

Kayodé, A. P. P., Bara, C. A., Dalodé-Vieira, G., Linnemann, A. R., & Nout, M. J. R. (2012). Extraction of antioxidant pigments from dye sorghum leaf sheaths. *LWT—Food Science and Technology*, *46*(1), 49–55. https://doi.org/10.1016/j.lwt.2011.11.003

Khan, M. I., & Giridhar, P. (2015). Plant betalains: chemistry and biochemistry. *Phytochemistry*, *117*, 267–295. https://doi.org/10.1016/j.phytochem.2015.06.008

Khatabi, O., Hanine, H., Elothmani, D., & Hasib, A. (2013). Extraction and determination of polyphenols and betalain pigments in the Moroccan prickly pear fruits (*Opuntia ficus indica*). *Arabian Journal of Chemistry*, in press. https://doi.org/10.1016/j.arabjc.2011.04.001

Kumar, S. S., Manoj, P., Shetty, N. P., Prakash, M., & Giridhar, P. (2015). Characterization of major betalain pigments – gomphrenin, betanin and isobetanin from *Basella rubra* L. fruit and evaluation of efficacy as a natural colourant in product (ice cream) development. *Journal of Food Science and Technology*, *52*(8), 4994–5002. https://doi.org/10.1007/s13197-014-1527-z

Lee, J. C., Kim, H. R., Kim, J., & Jang, Y. S. (2002). Antioxidant property of an ethanol extract of the stem of *Opuntia ficus-indica* var. saboten. *Journal of Agricultural and Food Chemistry*, *50*(22), 6490–6496. https://doi.org/10.1021/jf020388c

Loginova, K. V., Lebovka, N. I., & Vorobiev, E. (2011). Pulsed electric field assisted aqueous extraction of colorants from red beet. *Journal of Food Engineering*, *106*(2), 127–133. https://doi.org/10.1016/j.jfoodeng.2011.04.019

Madene, A., Jacquot, M., Scher, J., & Desobry, S. (2006). Flavour encapsulation and controlled release – a review. *International Journal of Food Science and Technology*, *41*(1), 1–21. https://doi.org/10.1111/j.1365-2621.2005.00980.x

Manchali S., Murthy K.N.C., Nagaraju S., Neelwarne B. (2013) Stability of betalain pigments of red beet. In: Neelwarne B. (eds) Red Beet Biotechnology. Springer, Boston, MA

Marañón-Ruíz, V. F., Rizo de la Torre, L. del C., & Chiu-Zarate, R. (2011). Caracterización de las propiedades ópticas de betacianinas y betaxantinas por espectroscopía UV-VIS y barrido en Z. *Superficies y Vacio*, *24*(4), 113–120.

Martín Villena, M. J., Morales Hernández, M. E., Gallardo Lara, V., & Ruiz Martínez, M. A. (2009). Técnicas de microencapsulación: una propuesta para microencapsular probióticos. *Ars Pharmaceutica*, *50*(1), 43–50.

Martínez, O. LA. (2015). Microencapsulación mediante secado por aspersión de compuestos bioactivos. *Revista Iberoamericana deTecnología Postcosecha*, *16*(2), 180–192. Retrieved from http://www.redalyc.org/pdf/813/81343176006.pdf

Robert de Mello, F., Bernardo, C., Odebrecht-Dias, C., Gonzaga, L., Amante, E.R., Fett, R., Bileski-Candido, L.M. (2015). Antioxidant properties, quantification and stability of betalains from pitaya (*Hylocereus undatus*) peel. *Ciencia Rural*, *45*(2), 323–328. https://www.scielo.br/pdf/cr/v45n2/0103-8478-cr-00-00-cr20140582.pdf

Moreno, M., Betancourt, M., Pitre, A., García, D., & Belén, D. (2007). Evaluación dela estabilidad de bebidas cítricas acondicionadas con dos fuentes naturales de betalaínas: tuna y remolacha. *Revista Facultad Nacional de Agronomía-Medellín*, *19*(3), 149–159.

Nava, C. G. (2010). *Caracterización fisicoquímica del fruto de garambullo (Myrtillocactus geometrizans)*. Master Dissertation, Universidad Autónoma de Querétaro, Mexico. https://doi.org/oai:ri.uaq.mx:123456789/2577.

Nilsson, T. (1970). Studies into the pigments in beet-root. *Lantbruckshogskolans Annaker*, *36*, 179–219.

Obón, J. M., Castellar, M. R., Alacid, M., & Fernández-López, J. A. (2009). Production of a red–purple food colorant from *Opuntia stricta* fruits by spray drying and its application in food model systems. *Journal of Food Engineering*, *90*(4), 471–479. https://doi.org/10.1016/j.jfoodeng.2008.07.013

Otálora, M. C., Carriazo, J. G., Iturriaga, L., Nazareno, M. A., & Osorio, C. (2015). Microencapsulation of betalains obtained from cactus fruit (*Opuntia ficus-indica*) by spray drying using cactus cladode mucilage and maltodextrin as encapsulating agents. *Food Chemistry*, *187*, 174–181. https://doi.org/10.1016/j.foodchem.2015.04.090

Özkan, G., & Bilek, S. E. (2014). Microencapsulation of natural food colourants. *International Journal of Nutrition and Food Sciences*, *3*(3), 145–156. https://doi.org/10.11648/j.ijnfs.20140303.13

Pangestuti, R., & Kim, S.-K. (2011). Biological activities and health benefit effects of natural pigments derived from marine algae. *Journal of Functional Foods*, *3*(4), 255–266. https://doi.org/10.1016/j.jff.2011.07.001

Parize, A.L., de Souza, T.C.R., Brighente, I.M.C., de Fávere, V.T., Laranjeira, M.C.M., Spinelli, A., Longo, E. (2008). Microencapsulation of the natural urucum pigment with chitosan by spray drying in different solvents. *African Journal of Biotechnology*, 7(17), 3107-3114. DOI:10.4314/ajb.v7i17.59236

Pérez-Alonso, C., Guadarrama-Lezama, A. Y., Barrera-Pichardo, J. F., Alamilla-Beltrán, L., & Rodríguez-Huezo, M. E. (2009). Interrelationship between the structural features and rehydration properties of spray dried manzano chilli sauce microcapsules. *Revista Mexicana de Ingeniería Química*, *8*(2), 187–196. Retrieved from http://www.redalyc.org/articulo.oa?id=62011384006

Pérez Leonard, H. (2006). Nutracéuticos: componente emergente para el beneficio de la Salud. *Instituto Cubano de Investigaciones de Los Derivados de La Caña de Azúcar*, *50* (3), 20–28.

Pérez-Leonard H., Bueno-García, G., Brizuela-Herrada, M. A., Tortoló-Cabañas, K.,& Gastón-Peña, C. (2013). Microencapsulación: una vía de protección para microorganismos probióticos. *Instituto Cubano de Investigaciones de Los Derivados de La Caña de Azúcar*, *47*(1), 14–25. Retrieved from http://www.redalyc.org/pdf/2231/223126409003.pdf

Pitalua, E., Jimenez, M., Vernon-Carter, E. J., & Beristain, C. I. (2010). Antioxidative activity of microcapsules with beetroot juice using gum Arabic as wall material. *Food and Bioproducts Processing*, *88*, 253–258. https://doi.org/10.1016/j.fbp.2010.01.002

Popplewell, L. M., Balck, J. M., Norris, L. M., Porzio, M., & Pszczola, D. E. (1995). Encapsulation system for flavors and colors. *Food Technology*, *49*(5), 76–82. Retrieved from http://cat.inist.fr/?aModele=afficheN&cpsidt=3522039

Puértolas, E., Cregenzán, O., Luengo, E., Álvarez, I., & Raso, J. (2013). Pulsed-electric-field-assisted extraction of anthocyanins from purple-fleshed potato. *Food Chemistry*, *136*(3–4), 1330–1336. https://doi.org/10.1016/j.foodchem.2012.09.080

Ravichandran, K., Palaniraj, R., Saw, N. M. M. T., Gabr, A. M. M., Ahmed, A. R., Knorr, D., & Smetanska, I. (2014). Effects of different encapsulation agents and drying process on stability of betalains extract. *Journal of Food Science and Technology*, *51*(9), 2216–2221. https://doi.org/10.1007/s13197-012-0728-6

Reynoso, R., Garcia, F. A., Morales, D., & Gonzalez De Mejia, E. (1997). Stability of betalain pigments from a cactacea fruit. *Journal of Agricultural and Food Chemistry*, *45*, 2884–2889. Retrieved from https://eurekamag.com/pdf.php?pdf=003279682

Sáenz, C., & Berger, H. (2006). *Utilización agroindustrial del nopal*. Organización de las Naciones Unidas para la Agricultura y la Alimentación, Roma. Retrieved from https://books.google.com/books?id=llaxlnmJjFoC&pgis=1

Saénz, C., Tapia, S., Chávez, J., & Robert, P. (2009). Microencapsulation by spray drying of bioactive compounds from cactus pear (*Opuntia ficus-indica*). *Food Chemistry*, *114*, 616–622. https://doi.org/10.1016/j.foodchem.2008.09.095

Sagdic, O., Ekici, L., Ozturk, I., Tekinay, T., Polat, B., Tastemur, B., ... Senturk, B. (2013). Cytotoxic and bioactive properties of different color tulip flowers and degradation kinetic of tulip flower anthocyanins. *Food and Chemical Toxicology: An International Journal Published for the British Industrial Biological Research Association*, *58*, 432–439. https://doi.org/10.1016/j.fct.2013.05.021

Salazar-González, C., Vergara-Balderas, F. T., & Guerrero-Beltrán, J. A. (2009). Evaluación de agentes antioxidantes de un extracto de flor de jamaica microencapsulado. *Temas Selec Tos de Ingenieria de Alimentos*, *3*, 14–25.

Sánchez-González, N. (2006). *Extracción y caracterización de los principales pigmentos del Opuntia joconoste c.v. (Xoconostle),* Master Dissertation, Center of Research in Applied Science and Advanced Technology, México.

Santos-Díaz, M. S., Velásquez-García, Y., & González-Chávez, M. M. (2005). Producción de pigmentos por callos de Mammillaria candida Scheidweiler (cactaceae). *Agrociencia*, *39*, 619–626. Retrieved from http://www.redalyc.org/articulo.oa?id=30239605

Scarafoni, A., Magni, C., & Duranti, M. (2007). Molecular nutraceutics as a mean to investigate the positive effects of legume seed proteins on human health. *Trends in Food Science & Technology*, *18*(9), 454–463. https://doi.org/10.1016/j.tifs.2007.04.002

Seigler, D. S. (2012). *Plant secondary metabolism*. Springer Science & Business Media, New York, USA. Retrieved from https://books.google.com/books?id=uKPwBwAAQBAJ&pgis=1

Olea-Serrano, M.F., López-Martínez, M.C., López- García de la Serrana, E. (2012). *Aspectos bromatológicos de conservantes y colorantes: toxicología alimentaria*. Ediciones Díaz de Santos, Madrid, España. Retrieved from https://books.google.com/books?id=g9FhUWsFY28C&pgis=1

Soriano-Santos, J., Franco-Zavaleta, M. E., Pelayo-Zaldívar, C., Armella-Villalpando, M.A.; Yáñez- López, M. L., & Guerrero-Legarreta, I. (2007). Caracterización parcial del pigmento rojo del fruto de la "Jiotilla" (*Escontria chiotilla* [Weber] Britton & Rose). *Revista Mexicana de Ingeníeria Química*, *6*(1), 19–25.

Stintzing, F. C., & Carle, R. (2004). Functional properties of anthocyanins and betalains in plants, food, and in human nutrition. *Trends in Food Science & Technology*, *15*(1), 19–38. https://doi.org/10.1016/J.TIFS.2003.07.004

Taper, H. S., Roberfroid, M. B. (1999). Nutritional and health benefits of inulin and oligofructose Influence of inulin and oligofructose on breast cancer and tumor growth 1. *The Journal of Nutrition*, 1488–1491. https://www.farm.ucl.ac.be/Full-texts-FARM/Taper-1999-2.pdf

Topete-Viniegra, R. (2006). Caracterización química y evaluación del efecto hipoglucemiante y antioxidante del fruto de Garambullo (*Myrtillocactus geometrizans*). Queretaro: Repositorio Institucional. Dirección de Innovación y Tecnologías de la Información. https://doi.org/oai:ri.uaq.mx:123456789/2160

Torres, V. (2008). *Microencapsulación de "polifenoles y betalaínas" desde un extracto acuoso de tuna (Opuntia ficus-indica), mediante secado por atomización.* Thesis, Universidad Tecnológica Metropolitana. México.

Ugaz, O. L. S. de. (1997). *Colorantes Naturales*. Fondo Editorial PUCP, Perú. Retrieved from https://books.google.com/books?id=LjmH_3qjaEIC&pgis=1

Vaillant, F., Perez, A., Davila, I., Dornier, M., Reynes, M., & Vaillant, F. (2005). Colorant and antioxidant properties of red-purple pitahaya (Hylocereus sp.). *Fruits*, *60*(601), 3–12. https://doi.org/10.1051/fruits:2005007

Vergara-Hinostroza, C. (2013). Extracción y estabilización de betalaínas de tuna púrpura (*Opuntia ficus-indica*) mediante tecnología de membranas y microencapsulación, como colorante alimentario, PhD Dissertation, Universidad de Chile, Chile. Retrieved from http://repositorio.uchile.cl/handle/2250/114868

CHAPTER 9

NANOEMULSIONS FOR EDIBLE COATINGS: STABILIZING AND BIOACTIVE PROPERTIES

MIGUEL DE LEÓN-ZAPATA[1], LORENZO PASTRANA-CASTRO[2], LETRICIA BARBOSA-PEREIRA[3], MARÍA L. RUA-RODRÍGUEZ[4], JANETH VENTURA[1], THALIA SALINAS[1], RAUL RODRÍGUEZ[1], JUAN A. ASCACIO-VALDÉS[1], JOSÉ SANDOVAL-CORTÉS[1], and CRISTÓBAL N. AGUILAR[1*]

[1]*Department of Research (DIA-UAdeC), School of Chemistry, University Autonomous of Coahuila. Saltillo, 25280 Coahuila, México*

[2]*Health, Food and Environment Department, International Iberian Nanotechnology Laboratory (INL), Braga 4715-330, Braga, Portugal*

[3]*Department of Agriculture, Forest and Food Science, School of Science, University of Turin, Turin 10095, Italy*

[4]*Laboratory of Analytical and Biochemistry Food, School of Science, University of Vigo, Ourense 32004, Galicia, España*

[*]*Corresponding author. E-mail: cristobal.aguilar@uadec.edu.mx*

ABSTRACT

The aim of this study is to evaluate the effect of extract of tarbush *Flourensia cernua* on the stabilizing, antioxidant, and fungistatic properties of a candelilla wax-based emulsion for edible coatings in refrigeration conditions for 7 weeks. The extract of tarbush was used as an active component of emulsions. The extract of tarbush presented good antioxidant activity in oil-in-water emulsion, as measured by the inhibition of the hydroperoxide assay by conjugated dienes, 2,2′-Azino-bis(3-ethylbenzothiazoline-6-sulfonic acid) diammonium salt (ABTS) and 2,2-diphenyl-1-picrylhydrazyl (DPPH). Results of the microbiological analysis demonstrated that the extract presented a higher fungistatic effect on yeast growth in the emulsion. The emulsion with the extract showed higher stability and antioxidant and fungistatic activity relative to control, without any significant differences in the storage for 4 weeks in refrigeration conditions.

9.1 INTRODUCTION

Emulsified systems are thermodynamically unstable due to the surface tension between oil and water, which opposes to the increase of interfacial area and can be stabilized by amphiphilic molecules, which adsorb at the oil–water interface, decreasing the surface tension between the two phases, called pickering emulsions (Figueroa-Espinoza et al., 2015).

The stability of emulsions is conditioned by the competition between attractive (van der Waals, hydrophobic interactions, electrostatic attractions, hydrogen bonds) and repulsive forces (electrostatic repulsion, steric repulsion) between the dispersed droplets (Guzeyand McClements, 2007) and depends on the constituents of the emulsion, namely, the concentration of the emulsifier or stabilizing agent, the pH, viscosity, and ionic strength of the aqueous phase, and the concentration of the organic phase (Tcholakova et al., 2006).

Dispersal of lipids in emulsified systems increases the specific area in contact with oxygen and some pro-oxidizing species (Coupland and McClements, 1996), which are detrimental to lipids.

One of the most effective and convenient strategies to retard or prevent lipid oxidation is to add antioxidants (Shahidi and Zhong, 2010). According to the mechanisms of action, antioxidants can be broadly classified as primary antioxidants, which scavenge free radicals to break chain reactions of oxidation, or secondary antioxidants, which protect lipids against oxidation mainly by chelating transition metals, quenching singlet oxygen, replenishing hydrogen to primary antioxidants, and/or scavenging oxygen (Reische et al., 2008).

Particularly, the regions of the north of Mexico, with its semiarid climate, have a great number and variety of wild plants grown under extreme climatic conditions; it is believed that some 25,000 species are registered and 30,000 not described (Adame and Adame, 2000).

One of them is tarbush (*Fluorensia cernua* D.C.), which is abundant in arid and semiarid regions of Mexico, where the tea brewed from the leaves of this plant is used in traditional medicine to treat digestive disorders, rheumatism, venereal diseases, herpes, bronchitis, varicella, and common cold (Ventura et al., 2009). It has been reported that the components of tarbush extracts have antioxidant andantifungal properties (De León-Zapata et al., 2016).

The biological activity of tarbush is due to its chemical composition mainly by compounds such as methyl orsellinate, ermanin, flourensadiol, dehydroflourensic acid, long-chain hydrocarbons from tetracosane 4-olide to triacontano-4-olide, and lactones (Jasso-De Rodríguez et al., 2007) in addition to saponins (Méndez et al., 2012), terpenes (Estell et al., 2013), condensed tannins equivalent to catechins (Méndez et al., 2012; De León-Zapata et al., 2013), and flavonoid glycosides (De León-Zapata et al., 2016).

Oil-in-water (O/W) emulsion is often more susceptible to oxidation than bulk oil due to its larger surface area that promotes interactions between the lipids and water-soluble prooxidants (Waraho et al., 2011). Oxidative reactions are believed to be most prevalent at the oil–water interface (McClements and Decker, 2000).

The high efficacy of antioxidants in O/W emulsions is primarily attributed to their high affinity to orient toward the oil–water interface (Waraho et al., 2011).

However, since lipid oxidation in O/W emulsions cannot be controlled by just the emulsifiers and thickeners, antioxidants are needed to further protect against rancidity, which may interact with antioxidants and affect the rate of oxidation. Moreover, the oxidative stability of structured lipids in real foods has seldom been investigated in the literature (Martin et al., 2010).

In this paper, the following properties of the extract of tarbush in a candelilla wax-based emulsion for edible coatings were studied: stability of the emulsion and fungistatic and antioxidant activity in the emulsified system during storage in refrigeration conditions.

9.2 MATERIALS AND METHODS

9.2.1 MATERIALS

Glicerol, arabic gum, Tween 80, and jojoba oil were supplied by Panreac (Madrid, Spain). Candelilla wax was supplied by Bioingenio Liftech S.A.

de C.V. (Saltillo, México). Ultrapure water was obtained from a Milli-Q filtration system (Millipore Corp., MA, USA).

9.2.2 VEGETAL MATERIAL

Leaves of tarbush *F. cernua* were collected from areas nearby to Saltillo, Coahuila, Mexico during March and April of 2015. The plant material was dehydrated at room temperature for 8–10 days and using a conventional oven (Labnet International, Inc.) at 60 ± 1 °C for 2 days. The leaves were stored in amber bottles or dark plastic bags at room temperature (25 ± 1 °C) until the obtention of the extract of tarbush.

9.2.3 PREPARATION OF EMULSION WITH THE EXTRACT OF TARBUSH

9.2.3.1 OBTAINING THE EXTRACT OF TARBUSH

The extracts of tarbush were obtained by an infusion method and heating reported by De León-Zapata et al. (2016). One sample of 10 g of leaves of tarbush was used and placed in an amber flask and then 100 mL of deionized water was added. The mixture was manually stirred and heated for 2 h at 60 ± 1 °C. The extract was filtered with a Wathman No. 1 paper, transferred to a glass Petri plate, and then placed in a conventional oven (Labnet International, Inc.) during 36 h at 60 ± 1 °C. The extracts of tarbush were stored in containers covered with aluminum or amber bottles at 5 ± 2 °C.

9.2.3.2 PREPARATION OF THE EMULSION

The emulsion was prepared using the hot high-shear stirring method (Solans et al., 2005). Briefly, Arabic gum (3% w/v) was homogenized using a high-shear stirrer Ultra-Turrax T25 Digital, IKA® (Staufen, Germany with an S25N-25 G, IKA disperser element) in distilled water at 800 rpm for 1 min and then heated to 85 ± 2 °C. Candelilla wax (1% w/v), jojoba oil (0.15%), glycerol (0.4%), and Tween 80 (0.8%) were added. For the emulsification of the components, a high-shear stirrer at 10,000 rpm for 5 min was used. A concentration of 3310 mg/L of the extract of tarbush was used in the emulsion based on the antifungal activity reported in a previous work (De León-Zapata et al., 2016). The samples were coded as EE (emulsion with extract) and control emulsion (emulsion without extract). These were characterized in terms of opacity and transparency and evaluated in terms of the stability of the system and fungistatic and antioxidant activity during storage.

9.2.4 OPACITY AND TRANSPARENCY

The opacity of the emulsion solutions was measured with a colorimeter (Model CR-400, Konica Minolta, Tokyo, Japan). The instrument was calibrated with a standard white plate (*Y*, *x*, *Y*). The measurements were performed in small Petri dishes, which contained 1 mL of liquid sample mounted on a plate. The opacity is determined by obtaining the CIE *Y* coordinates of the samples using five replicates, on a black and white background (Cerqueira et al., 2009) and is calculated as follows:

$$\text{Opacity} = Y_b/Y_w \times 100$$

where Y_b is the *Y* coordinate measured on the black background and Y_w is the *Y* coordinate measurement on the white background.

The transparencies of the emulsions at wavelengths ranging from 800 and 1000 nm were investigated (Salleh et al., 2009).

9.2.5 EVALUATION OF EMULSION SOLUTIONS

The EE and control emulsion were made to evaluate the effect of the extract of tarbush on the stability of the system and the fungistatic and antioxidant activity in the emulsified system during storage in refrigeration conditions (10 ± 1 °C). Periodic sampling was carried out each week during 7 weeks of storage in refrigeration conditions.

9.2.6 DETERMINATION OF STABILITY

The volumetric method was used to find out ESI for the emulsion stability of W/O emulsifiers (Lee et al., 2013). The emulsions were incubated at

100 ± 1 °C for 2 h when the separated layer was formed. All samples were measured in triplicate. ESI was calculated as follows:

$$ESI = \left[1 - \frac{\text{volume of separated layer}}{\text{total volume ofemulsion}}\right] \times 100$$

9.2.7 DETERMINATION OF ANTIOXIDANT ACTIVITY

The cation radical ABTS was synthesized by the reaction of a 7 mM ABTS solution with a 2.45 mM $K_2S_2O_8$ solution. The mixture was kept at 23±1 °C in the dark for 16 h. Afterward, the ABTS solution was diluted with ethanol until an absorbance of 0.7 at 734 nm was achieved using a UV–Vis spectrophotometer. In total, 10 µL of sample was added in the reaction cuvette, and immediately afterward, 1 mL of ABTS solution was added. After 10 min, the percentage inhibition of absorbance at 734 nm was calculated for each concentration, relative to the blank absorbance (ethanol). The DPPH radical is characterized by an unpaired electron, which is a free radical stabilized by resonance. A solution of DPPH radical at a concentration of 60 mM by diluting with methanol was prepared. In total, 100 µL of each sample of treatments was added in test tubes covered with foil and then 2.9 mL of DPPH solution was added and allowed to stand for 30 min. The absorbance was recorded at a wavelength of 517 nm. All samples were measured in triplicate.

The percentage inhibition of the radicals ABTS and DPPH was calculated as follows:

$$\text{Inhibition (\%)} = \frac{(A_{\text{control}} - A_{\text{sample}})}{A_{\text{control}}} \times 100$$

9.2.8 DETERMINATION OF CONJUGATED DIENE HYDROPEROXIDES

Hydroperoxides were determined by measuring conjugated dienes in 2-propanol at 234 nm using 26,000 as the molar extinction coefficient for methyl linoleate hydroperoxide (Schwarz et al., 2000). The results were expressed as millimole of hydroperoxides per kilogram of oil (mmol/kg oil), as described previously (Huang et al.,1996). All samples were measured by triplicate. % Inhibition was calculated as follows:

$$\% \text{ Inhibition} = ((C - S/C) \times 100$$

where C is the increment in the oxidation product formed in control and S is the increment in the oxidation product formed in sample, both expressed as mmol/kg oil.

9.2.9 *MICROBIOLOGICAL ASSAYS*

The count of UFC was done by the method of counting of fungi and yeasts in food based on the Official Mexican Norm NOM-111-SSA1-1994. Dilutions of 1:1000 of the EE and control emulsion in sterile phosphate solution were used. One milliliter of each sample was transferred and distributed using five replicates in Petri dishes. To each inoculated dish, approximately 15 mL of acidified potato dextrose agar with sterile tartaric acid was added at 45 ± 1 °C. The samples were mixed immediately after pouring by rotating the Petri dish sufficiently to obtain evenly dispersed colonies after incubation. After complete solidification, the plates were inverted and incubated at 25 ± 1 °C for 5 days. The count was expressed as CFU/mL of emulsion. All samples were measured in triplicate.

9.2.10 *STATISTICAL ANALYSIS*

The results were statistically evaluated by analysis of variance (ANOVA) and Tukey's test at the 5% significance level, using the software Statistica® 7 (StatSoft Inc., Tulsa, USA).

9.3 RESULTS AND DISCUSSION

Table 9.1 shows the effect of extract of tarbush on the opacity and transmittance of the emulsion at wavelengths of 800 and 1000 nm stored during 4 weeks in refrigeration conditions. The opacity and transmittance of the emulsions were assessed until the fourth week of storage in refrigeration since the control emulsion became unacceptable on the appearance after 4 weeks. The EE emulsion (Table 9.1) showed higher transmittance (88% and 97% at 800 and 1000 nm, respectively) and minor opacity (17%) with respect to the control emulsion (Table 9.1).

TABLE 9.1 Effect of the Extract of Tarbush on the Opacity and Transmittance of the Candelilla Wax-Based Emulsion Stored for 4 Weeks in Refrigeration Conditions (10 ± 1 °C)

Parameter	Treatments	
	EE	Control emulsion
Opacity (%)	17 a	48 b
Transmitance (%) to 800 nm	88 a	80 b
Transmitance (%) to 1000 nm	97 a	89 b

EE: Emulsion with the extract of tarbush. Within a row, different letters represent a significant difference ($P < 0.05$).

The presence of the extract of tarbush in the emulsion decreased the opacity and increased the transparency, which is evident in the appearance of the emulsions (Figure 9.1). The transmittance of the emulsion was higher with the extract of tarbush (Table 9.1). This is because the extract reduces interactions between polymer molecules, which results in a structure with visible cracks (Chen et al., 2009) through which light passes easily, thereby decreasing the opacity of the system (Carneiro Da Cunha et al., 2009).

Control emulsion EE

FIGURE 9.1 EE and control emulsion stored for 4 weeks in refrigeration conditions.

Transmittance is directly correlated to the particle size (Salleh et al., 2009), which it is attributed to the fact that small particles scatter light weakly; therefore, as the particles size increases, the light scattering becomes strong and emulsions tend to be opaque (Acevedo-Fani et al., 2015).

Even though the microbial analysis accounts for both fungi and yeast, visual analysis of the plates indicated that most of the microorganisms in the emulsions were yeast (Figure 9.2). Yeast growth in the control emulsion began in the second week (12 CFU/mL) and continued until the seventh week (47 CFU/mL) of storage (Figure 9.2). The extract of tarbush managed to inhibit the growth of fungi and yeasts in the system until the fourth week of refrigerated storage (Figure 9.2). Yeast growth in the EE began in the fifth week (2 CFU/mL) and continued until the seventh week (8 CFU/mL) of storage (Figure 9.2) due to the fungistatic effect of the extract. The comparison of growth rates in the emulsions indicated that EE had lower values of CFU than the control emulsion (Figure 9.2).

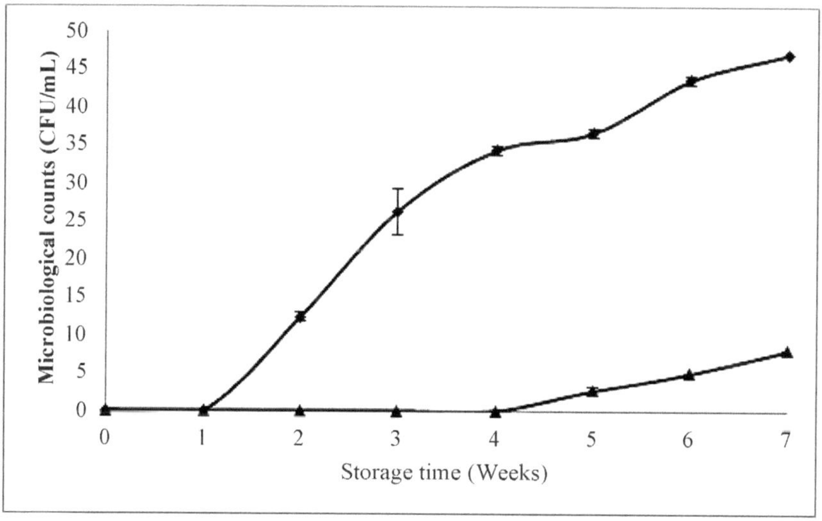

FIGURE 9.2 Microbial growth in the EE (▲) and control emulsion (♦) stored for 7 weeks in refrigeration conditions.

The concentration used of the extract of tarbush in the emulsion (3,310 mg/L) contains from 4.24 to 5.80 mg of gallic acid and glucosides of flavonoids such as luteolin 7-O-rutinoside and apigenin galactoside arabinoside with antioxidant activity (6.07–7.62 μmol/g of TEAC) and antifungal activity against *Rhizopus stolonifer*, *Botrytis cinerea*, *Fusarium Oxysporum*, and *Colletrotrichum gloeosporioides* (De León-Zapata et al., 2016).

The greater antioxidant (Table 9.2) and fungistatic (Figure 9.2) effectiveness of the extract of tarbush is attributable to the phenolic compounds present as hydroxyl groups of hydrolyzable tannins equivalent to gallic acid and glucosides of flavonoids (De León-Zapata et al., 2016).

These compounds have the ability to form complexes with proteins and polysaccharides of the microorganism, inhibiting the electron transport through membranes (Scalbert and Williamson, 2000) and causing cell lysis (Cowan, 1999).

TABLE 9.2 Effect of the Extract of Tarbush on the Antioxidant Activity of the Candelilla Wax-Based Emulsion Stored for 7 weeks in Refrigeration Conditions (10 ± 1 °C).

Storage time (weeks) in refrigeration		**0**	**2**	**4**	**5**	**6**	**7**
Control emulsion	ABTS (%)	5 a	7 a	9 a	1 b	0 b	0 b
	DPPH (%)	0 a	0 a	0 a	0 a	0 a	0 a
	Inhibition of hydroperoxides (%)	2 a	4 a	5 a	0 b	0 b	0 b
EE	ABTS (%)	53 a	55 a	56 a	30 b	24 b	22 b
	DPPH (%)	46 a	48 a	50 a	20 b	16 b	12 b
	Inhibition of hydroperoxides (%)	86 a	88 a	89 a	43 b	34 b	32 b

EE: Emulsion with the extract of tarbush. Within a row, different letters represent a significant difference ($P < 0.05$).

Tannins and flavonoids contain in their chemical structure a variable number of hydroxyl groups (Martínez-Flórez et al., 2002), which are involved in neutralizing free radicals by donating electrons and thus influence the antioxidant activity (De León-Zapata et al., 2013).

The values obtained by the capture of the ABTS radical are higher than those obtained by the DPPH radical (Table 9.1) due to the sensitivity of the ABTS radical because it is a structure that easily reacts with hydrophilic and lipophilic compounds (Pérez et al., 2003) and reducing agents (Rojano et al., 2009). However, the DPPH radical reacts with hydrophilic compounds such as gallic acid and flavonoids (Álvarez et al., 2008).

The results of the antioxidant activity by ABTS and DPPH are consistent with those obtained for the inhibition of hydroperoxides in the emulsion (89%), which remained stable until the fourth week of refrigerated storage, protecting the emulsified system from lipid oxidation (Table 9.2) due to the neutralization of free radicals by a large number of hydroxyl groups present in the extract of tarbush, which remain between the aqueous and oil phases of the emulsion.

The antioxidant activity of the extract of tarbush is mainly due to its redox properties, which play an important role in the absorption and neutralization of free radicals (De León-Zapata et al., 2013). Its antioxidant potential depends on the number of hydroxyl groups and the conjugation degree of the structure because the activity improves when the number of hydroxyl groups increases (Sang et al., 2002) such as the gallic acid and flavonoids, which are phenolic compound with OH groups and carboxylic acids (Méndez et al., 2012).

The results obtained are related to the part of stability of the emulsion (Figure 9.3) because lipid oxidation can be retarded by strengthening the interfacial layer and by adsorbing tannins in the oil and aqueous phases, which may affect oxygen transfer and oxidation products (Ma et al., 2012).

Figure 3 shows that the absence of the extract of trabush in the emulsified system (control emulsion) promoted a lower stability index (93.84%–93.63%) during storage under refrigeration.

The EE showed a higher index stability during refrigerated storage, where the stability is stable until four weeks (Figure 9.3).

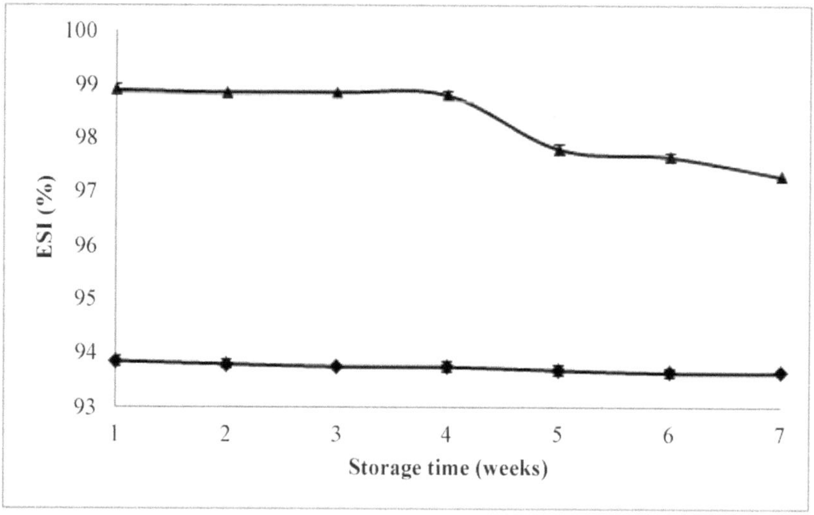

FIGURE 9.3 Effect of time (weeks) on the stability of the EE (▲) and control emulsion (♦) stored for 7 weeks in refrigeration conditions.

The extract of tarbush provides greater stability to the emulsion (Figure 9.3) due to content of tannin and glycosides of flavonoids (De León-Zapata et al., 2016), which are adsorbed on the surface of oil globules present in the emulsion and are very difficult to remove once adsorbed at the interface, making it a more stable emulsion, due to steric and/or electrostatic repulsions between the droplets covered by adsorbed tannins (Figueroa-Espinoza et al., 2015).

When the polymerization degree of the tannins adsorbed is higher, the stability of the emulsion increases due to the formation of thicker layers, leading to stronger steric repulsion between the droplets and preventing their coalescence or flocculation (Figueroa-Espinoza et al., 2015).

These results agree with those reported by Lee et al. (2006), who observed that antioxidants such as all-*trans*-retinol improve the stability of solid lipid nanoparticles.

From the fifth week, a decrease was observed in the stability of EE during storage. The decreased of stability (Figure 9.3) may be related to the results of microbiological analysis because in the fifth week of storage yeast growth was observed in the system.

The ability of these microbes to penetrate into the oil–water interface was related to their cell surface properties, as lactic acid bacteria and yeasts, which live and grow in aqueous environments, naturally show a predominantly hydrophilic behaviour (Firoozmand and Rousseau, 2016). This type of behavior shows that yeasts have the capacity to grow in the aqueous phase of the emulsion and consume the available sugars of glucosides present in the phytomolecules of tarbush, therefore destabilizing the emulsified system by interfering with the aqueous and oil phases of the emulsion. This is largely due to the dominance of carbohydrates in yeast cell walls (94%) (Dallies et al., 1998). Yet, yeasts have been shown to attach at the oil–water interface via hydrophobic interactions (Dorobantu et al., 2004), with such behaviour broadly determined by the composition and conformation of surface-bound proteins, polypeptides, and polysaccharides (Chapot-Chartier and Kulakauskas, 2014).

Based on the above, possibly, the yeasts interacted with the aqueous phase of the emulsion and may have consumed sugars present in the glucosides of flavonoids of the extract of tarbush, thus destabilizing the emulsified system between the aqueous and oil phases and affecting the stability and antioxidant (Table 9.2) and fungistatic activity of the extract in the emulsion until the fifth week of storage in refrigeration.

9.4 CONCLUSIONS

This is the first report in publicize the use of the extract of tarbush as a fungistatic, an antioxidant, and a stabilizer in a candelilla wax based-emulsion (O/W). Emulsion stabilization would be due to the steric repulsions and/or stabilization by tannin aggregates. The extract of tarbush provided higher transparency to the emulsion. It was demonstrated, for the first time, that the extract of tarbush presented good antioxidant and fungistatic activity in O/W emulsion, and thus it protects oil against oxygen damages. This research shows the potential use of high added-value extract of tarbush distributed in semi-arid regions of Mexico that could be used as an antioxidant emulsifier for potential use as a natural stabilizer of emulsions.

ACKNOWLEDGMENTS

The authors thank the financial support provided by the National Council of Science and Technology (CONACYT), Autonomous University of Coahuila (UAdeC), and BIOINGENIO LIFETECH S.A. de C.V.

KEYWORDS

- **candelilla**
- **tarbush**
- **emulsion**
- **stabilizers**
- **antioxidants**

REFERENCES

Acevedo-Fani, A., Salvia-Trujillo, L., Rojas-Graü, M.A., and Martín-Belloso, O. (2015). Edible films from essential-oil-loaded nanoemulsions: Physicochemical characterization and antimicrobial properties. *Food Hydrocolloids*, *47*, 168–177.

Adame, J., and Adame, H. (2000). *Plantas curativas del Noreste Mexicano (Ediciones Castillo) Monterrey* (pp. 11–15). México: Nuevo León.

Álvarez-R., E., Jimenez-G., O.J., Posada-A., C.M., Rojano, B.A., Gil-G., G.H., García-P., C.M., Durango-R., D. L. (2008). Actividad antioxidante y contenido fenólico de los extractos provenientes de las bayas de dos especies del género Vismia (Guttiferae). *Vitae, Revista de la Facultad de Química Farmacéutica, 15*, 165–172.

Carneiro Da Cunha, M.G., Cerqueira, M.A., Souza, B.W.S., Souza, M.P., Teixeira, J.A., and Vicente, A.A. (2009). Physical properties of edible coatings and films made with a polysaccharide from *Anacardiumoccidentale* L. *Journal of Food Engineering, 95*, 379–385.

Cerqueira, M.A., Lima, A.M.P., Teixeira, J.A., Moreira, R.A., and Vicente, A. A.(2009). Suitability of novel galactomannans as edible coatings for tropical fruits. *Journal of Food Engineering, 94*, 372–378.

Chapot-Chartier, M.P., and Kulakauskas, S. (2014). Cell wall structure and function in lactic acid bacteria. *Microbial Cell Factories, 13*, 1–23.

Chen, C.H., Kuo, W.S., and Lai, L. S. (2009). Effect of surfactants on water barrier and physical properties of tapioca starch/decolorized hsian-tsao leaf gum films. *Food Hydrocolloids, 23*, 714–721.

Coupland, J.N., and McClements, D. J. (1996). Lipid oxidation in food emulsions. *Trends in Food Science and Technology, 7*, 83–91.

Cowan, M. M. (1999). Plant products as antimicrobial agents. *Clinical Microbiology Reviews, 12*, 564–582.

Dallies, N., François, J., and Paquet, V. (1998). A new method for quantitative determination of polysaccharides in the yeast cellwall. Application to the cell wall defective mutants of *Saccharomyces cerevisiae*. *Yeast, 14*, 1297–1306.

De León-Zapata, M.A., Pastrana-Castro, L., Rua-Rodríguez, M.L., Alvarez-Pérez, O.B., Rodríguez-Herrera, R., Aguilar, C.N. (2016). Experimental protocol for the recovery and evaluation of bioactive compounds of tarbush against postharvest fruit fungi. *Food Chemistry, 198*, 62–67.

De León-Zapata, M.A., Sáenz, A., Jasso-Cantu, D., Rodríguez, R., Pandey, A., and Aguilar, C. N. (2013). Fermented *Flourensiacernua* extracts and their in vitro assay against *Penicillium expansum* and *Fusarium oxysporum*. *Food Technology and Biotechnology, 51*, 233–239.

Dorobantu, L.S., Yeung, A.K.C., Foght, J.M., and Gray, M. R. (2004). Stabilization of oil–water emulsions by hydrophobic bacteria. *Applied and Environmental Microbiology, 70*, 6333–6336.

Estell, R.E., James, D.K., Fredrickson, E.L., and Anderson, D.M.(2013). Within-plant distribution of volatile compounds on the leaf surface of *Flourensiacernua*. *Biochemical Systematics and Ecology, 48*, 144–150.

Figueroa-Espinoza, M.C., Zafimahova, A., Maldonado-Alvarado, P., Dubreucq, E., and Poncet-Legrand, C. (2015). Grape seed and apple tannins: Emulsifying and antioxidant properties. *Food Chemistry, 178*, 38–44.

Firoozmand, H., and Rousseau, D. (2016). Microbial cells as colloidal particles: Pickering oil-in-water emulsions stabilized by bacteria and yeast. *Food Research International, 81*, 66–73.

Guzey, D., and McClements, D. J. (2007). Impact of electrostatic interactions on formation and stability of emulsions containing oil droplets coated by blactoglobulin–pectin complex. *Journal of Agricultural and Food Chemistry, 55*, 475–485.

Huang, S.W., Frankel, E.N., Schwarz, K., Aeschbach, R., and German, J. B. (1996). Antioxidant activity of carnosic acid and methyl carnosate in bulk oils and oilin-water emulsions. *Journal of Agricultural and Food Chemistry*, *44*, 2951–2956.

Jasso-De Rodríguez, D., Hernández, C.D., Angulo, S.J.L., Rodríguez, G.R., Villarreal, Q.J.A., and Lira, S.R.H. (2007). Antifungal activity in vitro of *F. cernua* extracts on *Alternaria* sp., *Rhizoctoniasolani*, and *Fusarium oxysporum*. *Industrial Crops and Products*, *25*, 111–116.

Lee, Y.K., Ahn, S., and Kwak, H.S. (2013). Optimizing microencapsulation of peanut sprout extract by response surface methodology. *Food Hydrocolloids*, *30*, 307–314.

Lee, J.P., Lim, S.J., Park, J.S., and Kim, Ch. K. (2006). Stabilization of all-*trans* retinol by loading lipophilic antioxidants in solid lipid nanoparticles. *European Journal of Pharmaceutics and Biopharmaceutics*, *63*, 134–139.

Ma, H.R., Forssell, P., Kylli, P., Lampi, A.M., Buchert, J., and Boer, H. (2012). Transglutaminase catalyzed cross-linking of sodium caseinate improves oxidative stability of flaxseed oil emulsion. *Journal of Agricultural and Food Chemistry*, *60*, 6223–6229.

Martin, D., Reglero, G., and Senorans, F. J. (2010). Oxidative stability of structured lipids. *European Food Research and Technology*, *231*, 635–653.

Martínez-Flórez, S., González-Gallegos, J., Culebras, J.M., and Tuñón, M. J. (2002). Los flavonoides: Propiedades y acciones antioxidantes. *Nutricion Hospitalaria*, *17*, 271–278.

McClements, D.J., and Decker, E. A. (2000). Lipid oxidation in oil-in-water emulsions: Impact of molecular environment on chemical reactions in heterogeneous food systems. *Journal of Food Science*, *65*, 1270–1282.

Méndez, M., Rodríguez, R., Ruiz, J., Morales-Adame, D., Hernández-Castillo, F.D., and Aguilar, C. N. (2012). Antibacterial activity of plant extracts obtained with alternative organic solvents against food-borne pathogen bacteria. *Industrial Crops and Products*, *37*, 445–450.

Pérez, R.M., Vargas, R., Martinez, F.J., Garcia, E. V., and Hernández, B. (2003). Actividad antioxidante de los alcaloides de Bocconiaarborea. Estudio sobre seis métodos de análisis. *Ars Pharmaceutica*, *44*, 5–21.

Reische, D.W., Lillard, D.A., and Eitenmiller, R. R. (2008). Antioxidants. In: C.C., Akoh, D.B., Min. (Eds.), *Food lipids: Chemistry, nutrition, and biotechnology* (pp. 409–430). Boca Raton, FL: CRC Press.

Rojano, A., Gaviria, C.A., Ochoa, C.I., Sánchez, N., Medina, C., and Lobo, M. (2009). Propiedadesantioxidantes de los frutos de agraz o mortiño (*Vaccinium meridionale* Swartz). In: *Perspectivas del cultivo de agraz o mortiño en la zona altoandina de Colombia* (pp. 95–112). Bogotá, Colombia: Gente Nueva Editorial.

Salleh, E., Muhamad, I.I., and Khairuddin, N. (2009). Structural characterization and physical properties of antimicrobial (AM) starch-based films. *World Academy of Science, Engineering and Technology*, *55*, 432–440.

Sang, S., Lapsley, K., Jeong, W.S., Lachance, P.A., Ho, C.T., and Rosen, R. T. (2002). Antioxidative phenolic compounds isolated from almond (*Prunus amygdalus* Batsch). *Journal of the Science of Food and Agriculture*, *50*, 2459–2463.

Scalbert, A., and Williamson, G. (2000). Dietary intake and bioavailability of polyphenols. *Journal of Nutrition*, *130*, 2073–2085.

Schwarz, K., Huang, S.W., German, J.B., Tiersch, B., Hartmann, J., andFrankel, E. N. (2000). Activities of antioxidants are affected by colloidal properties of oil in water and water in oil emulsions and bulk oils. *Journal of Agricultural Chemistry*, *48*, 4874–4882.

Shahidi, F., and Zhong, Y.(2010). Lipid oxidation and improving the oxidative stability. *Chemical Society Reviews*, *39*, 4067–4079.

Solans, C., Izquierdo, P., Nolla, J., Azemar, N., and Garcia-Celma, M. J. (2005). Nano-emulsions. *Current Opinion in Colloid and Interface Science*, *10*, 102–110.

Tcholakova, S., Denkov, N.D., Sidzhakova, D., and Campbell, B. (2006). Effect of thermal treatment, ionic strength, and pH on the short-term and long-term coalescence stability of beta-lactoglobulin emulsions. *Langmuir*, *22*, 6042–6052.

Ventura, J., Gutiérrez-Sánchez, G., Rodríguez-Herrera, R., and Aguilar, C. N. (2009). Fungal cultures of tarbush and creosote bush for production of two phenolic antioxidants (pyrocatechol and gallic acid). *Folia Microbiologica*, *54*, 199–203.

Waraho, T., McClements, D.J., and Decker, E.A. (2011). Mechanisms of lipid oxidation in food dispersions. *Trends in Food Science and Technology*, *22*, 3–13.

CHAPTER 10

ENHANCING THE ADDED VALUE OF SORGHUM BY BIOMOLECULE CONTENT AND BIOPROCESSING

MARISOL CRUZ-REQUENA[1], LEOPOLDO J. RÍOS-GONZÁLEZ[2], JOSÉ ANTONIO DE LA GARZA-RODRÍGUEZ[2], SÓCRATES PALACIOS PONCE[3], and MIGUEL A. MEDINA-MORALES[2*]

[1]*Food Research Department, School of Chemistry, Autonomous University of Coahuila, Saltillo 25280, Coahuila, México*

[2]*Biotechnology Department, School of Chemistry, Autonomous University of Coahuila, Saltillo 25280, Coahuila, México*

[3]*Mechanical Engineering and Production Sciences School, ESPOL Polytechnic University (Litoral), Campus Gustavo Galindo, CP 09-01-5683 Guayaquil, Ecuador*

[*]*Corresponding author. E-mail: miguel.medina@uadec.edu.mx*

ABSTRACT

Sorghum is a relatively undervalued crop compared to corn and wheat. As it is a plant that thrives better in dry environments and has a versatile carbon fixation metabolism, it may represent an important underused source of high added-value chemicals and food. By biotechnological processing, several outcomes may prove said importance of sorghum. Most of the plant is useful for processing by biotechnological or by food bioscience way. For these purposes, the content of bioactive molecules, such as polyphenols as relative majority and other compounds such as stilbenoids. It is important to add that, for most of the native bioactive content of sorghum, physical processing may facilitate the assimilation of those molecules in the human diet with many beneficial effects. Also, the polysaccharide content can be processed for bioactive or high added-value molecules production by the biotechnological way. By enzymatic processing, starch degree of polymerization can be disrupted by liquefaction, and by microbial metabolism, several products may be produced. It is worth noting that the compounds mentioned in this chapter have application in many industrial areas.

10.1 INTRODUCTION

Human good quality of life is the most prominent area of research, which goes hand by hand with food technology and biotechnology (Garaguso and Nardini, 2015). It is widely known that the way we eat, has great repercussions in our lifestyle and well-being. In our diet, there are food sources that contain several compounds that show bioactivities that help improve health in humans (Singh et al., 2019). The sources considered are from plants, such as grains (Lima et al., 2019). From these, also the leaves and stalks, which most of the time are residues or wastes, represent the raw material for bioactive and added value compounds (Rojas et al., 2018). There are several types of compounds that are extractable and/or obtainable from these materials, and the molecules that possess an interest in fuel, pharmaceutical, food, and cosmetic industries are bioalcohols and polyphenols (Dey and Kuhad, 2014). These molecules are a group of metabolites that can be subdivided into several categories, of which the most acknowledged are phenolic acids, stilbenes, lignans, and flavonoids (Kumar et al., 2019; Majdalawieh and Mansour, 2019). Many of the compounds that are present in plant tissues are flavonoids, which are responsible for color and defense mechanisms (Mark et al., 2019). Most of the flavonoids and phenolic acids have been studied for their antioxidant, antimicrobial, antitumoral, and other bioactivities, and their inclusion in food products is a practice that nowadays is common for the previously stated reasons (do Prado et al., 2014). Flavonoids include apigenin, luteolin, catechin, and kaempferol, and phenolic acids include cinnamic and benzoic acid derivates (López-Trujillo et al., 2017; Sepúlveda et al., 2016). Sorghum (*Sorghum bicolor*) is a very versatile crop because it is a C_4 carbon fixation plant, which converts CO_2 into carbohydrates. This type of metabolism makes sorghum highly resistant to drought, and it can not only grow on drylands but also proliferate in tropical, subtropical, and cooler environments (Gomes et al., 2019). Mexico is the third largest producer of sorghum in the world, which gives an advantage of production in that regard. There are high numbers of sorghum production but is mostly used as animal feed. Its content of phenolic acids and flavonoids and its starch type make sorghum an attractive food source, which may turn as functional for the same reasons. In Africa, Western Europe, India, and China, it is very common to find flatbreads, snacks, fermented foods, and alcoholic and nonalcoholic beverages. There are studies where sorghum extracts have antagonist activities against esophageal cancer and chronic diseases (Luo et al., 2018). From another perspective, the sugars available in sorghum are another alternative for increasing the crop value. Biotechnology allows the possibility to obtain metabolites by microbial processing of sugar and its subsequent biotransformation. Firstly, several added-value compounds produced from grains could be considered nonviable because it could affect the basic food product consumers who buy in markets. Corn, for instance, is considered commercially nonviable as a resource for ethanol production. Alternatively, other grains can be used for the same purpose, such as sorghum (Lolasi et al., 2018). From the grains, starch can be used for microbial metabolite production via enzymatic hydrolysis and fermentation. According to these facts, sorghum sugar content can be considered as a viable feedstock from the starch of the grains or the lignocellulosic material from stalks and leaves. In the case of grains, amylolytic enzymes are required to degrade starch. Enzymes such as α-amylase and pullulanase are required to release glucose from the polysaccharides for microorganisms to generate the products of interest (Lolasi et al., 2018). Second, the stalks also provide a substrate for production, where sweet sorghum stalks accumulate sugars similarly to sugarcane; therefore, direct production of metabolites can be achieved from stalk juice. If higher amounts are needed or desired to produce, the lignocellulosic fraction of the sorghum plant can also be processed for it. In this case, a more complex process has to be carried out to pretreat the material, enzymatically degrade cellulose and hemicellulose to monomeric sugars (Dar et al., 2018; Arenas-Cárdenas et al., 2016). Once the release and accumulation of sugars are reached, microbial production is the next step. For microbial metabolite production, several microorganisms have been used, such as *Saccharomyces cerevisiae*, *Zymomonas mobilis*, or *Kluyveromyces marxianus* for ethanol (Verdugo-Valdez et al., 2011), *Debaryomyces hansenii* and *Candida guillermondii* for xylitol (Khlestkin et al., 2018), Lactobacilli for lactic acid and short-chain fatty acids (Jagtap and Rao, 2018), and Clostridia for microbial solvents, which can be further processed by lipases to produce butyl butyrate (Xin et al., 2019). After considering that sorghum has potential in its processing, not only for animal feed, as it is the most common use, this chapter will cover a number of extraction technologies considered as emergent used on sorghum to extract phytochemicals. Biotechnological processing of polysaccharides and sugars contained in the plant to obtain added-value compounds will be mentioned. Also, the nutritional value of sorghum will be addressed in this document.

10.2 SORGHUM

Sorghum bicolor is a versatile and drought-tolerant crop commonly produced in semi-arid regions of Africa, Asia, Australia, and North and South America (Ahmad et al., 2018). The varieties of grain sorghum have a similar starch content to that of maize (60%–77%), making this crop an alternative for ethanol production, particularly in regions with low precipitation, where corn cultivation is complicated (Impa et al., 2019). As a C_4 photosynthetic plant, sorghum is highly efficient at assimilating carbon and accumulating biomass at high temperatures, in contrast to C_3 cereal plants such as rice and wheat. In addition to grain production, sorghum is also grown for fodder, sugar (sweet stalk), and second-generation biofuels (through lignocellulosic biomass) (Girard and Awika, 2018). This variety of sorghum is considered more cost-efficient compared to corn due to low nitrogen requirements and water culture. The composition of the stems varies according to the genetic, climatic, and edaphic factors (Ahmad et al., 2018). On average, the stems are composed of 43%–58% of soluble substances (mainly sucrose, glucose, and fructose) and 22.6%–47.8% of insoluble substances (mainly cellulose and hemicellulose). However, you can also find other sugars such as arabinose, galactose, mannose, and xylose (Kaur et al., 2010). The maximum sugar content (16–23 °Brix) can be found in the prebloom, decreasing significantly (20%–25%) when harvested after this stage (Amaducci et al., 2004).

10.2.1 NUTRITIVE VALUE OF SORGHUM GRAINS

One of the most important cereals in the world is sorghum since it is the fifth most produced worldwide after rice, wheat, corn, and barley and is amply used for bioethanol production as well as human and animal consumption (Ramatouyale et al., 2016). The nutritive value of this cereal varies according to the variety, but in general, these grains contain low-digestible protein and starch, polyphenols (condensed tannins), dietary fiber, fat-soluble vitamins and vitamin B, and minerals. The interest of the consumption of this grain lies in its indisputable nutritive value, which has been studied amply, and, on the other side, sorghum is a gluten-free cereal, which means that it can be consumed by patients with celiac disease (Zhu, 2014). As food, sorghum is consumed in Asia and Africa in different forms like tortilla, porridges, couscous, baked goods, snacks, cookies, bread, etc. (Kulamarva et al., 2009). However, its consumption in other parts of the world consumption is common, in the American continent it is not and there have been proposals to increase sorghum consumption due to its nutritional and functional properties as food or as a functional ingredient in several food for its positive impact on human health (Stefoska-Needham et al., 2015). Likewise, sorghum grains are a source of two types of dietary fiber: insoluble fibers, principally arabinoxylans, and soluble fibers with a presence in the grain of 10%–20% (Morais et al., 2015). Arabinoxylans are the principal dietary fibers found in cereals and have been documented for their impact on the host immunity, enhancing innate immunity and protecting against chronic inflammatory diseases (Fadel et al., 2018).

10.2.2 SORGHUM AND DIETARY FIBER

The content of dietary fiber in sorghum grains is represented mainly by resistant starch (Birt et al., 2013). Resistant starch is a type of starch that is resistant to digestive enzymes and passes through the upper digestive tract to the colon, where they are fermented by microbiota, producing short-chain fatty acids, principally butyrate (Shen et al., 2015). In this sense, the resistant starch in sorghum grains has been studied and one of its functional properties lies in the reduction of body fat and the improvement of lipid metabolism disorder; also, the sorghum-resistant starch can improve the intestinal microbiota (Shen et al., 2015).

10.2.3 MINERAL CONTENT

The mineral content of sorghum grains has placed it as a potential source and an effective component of functional foods and improved food nutritional quality. Among minerals in these grains, copper, zinc, iron, manganese, calcium, and magnesium can be found (Samarth et al., 2018). Furthermore, some processes such as germination, decortication and/or malting, and fermentation enhance the nutritional value or this cereal, changing and enhancing the chemical composition and eliminating antinatural factors, including minerals; for example, iron is not available in the presence of polyphenols and phytates, but after fermentation to obtain beer, the bioavailability of this mineral increase (Léder, 2004).

10.2.4 USEFUL COMPONENTS OF SORGHUM

Nowadays, sorghum grains have been used mostly as a feed crop, but the interest for this cereal has been increasing because of the search for sources rich in bioactive compounds as potential ingredients in functional foods. The sorghum grain by itself has a high nutritional value since it contains low-digestibility proteins, unsaturated lipids, minerals, vitamins, and phenolic compounds, especially 3-deoxyanthocyanidins and tannins (Morais et al., 2015). Grains are base of several food products from bakery to beer, and for their production, the cereal grains goes through various physicochemical changes. One of the food procedures used in cereal is extrusion, which modifies the nutritional content and therefore the functional properties of sorghum grains. One of the principal changes is the content of phenolic acid compounds after the extrusion process due to the increase of the retention of low-molecular-weight proanthocyanidins, favoring higher bioavailability of these compounds and hence greater efficacy against oxidative stress (Rodrigues de Souza et al., 2019).

There are several scientific studies that mention the functional properties of sorghum grains as well as the health benefits that are obtained after their consumption. On the other side, starch present in the endosperm has a slow digesting profile compared to other cereal grains, which has shown to modulate the postprandial blood glucose response in humans (Simnadis et al., 2016; Vila-Real et al., 2017). In most, if not all, of the plant tissues in nature, bioactive compounds are present. The function of the molecules is mainly plant tissue protection against UV radiation and from predators and microorganisms (Makris, 2018).

The intention in this case is the possibility of the use of sorghum grain as a source of polyphenols, as it is one of the agricultural products produced majorly. Several plant materials are reported as a source of polyphenols such as pecan nuts, pomegranate, mango, and citrus fruits (Medina et al., 2010; Buenrostro-Figueroa et al., 2013; Rojas et al., 2018). In light of this, the most common polyphenolic components in plants are phenolic acids, flavonoids, and stilbenoids (Bröhan et al., 2011). One of the principal bioactive components of the sorghum grain are polyphenols, which are amply studied and have demonstrated their antioxidant capacity. Among these phenolic compounds are phenolic acids, flavonoids, and condensed tannins, concentrated in the pericarp, and the level and composition of these are affected by the genotype (Aruna and Visarada, 2019). Sorghum grains present a diversity of polyphenols according to the color of the cereal, and these colors range from cream white to red, lemon yellow, and black. The principal phenolic acids present in sorghum grains are cinnamic and benzoic derivatives, and they are bonded to hemicelluloses. On the other side, the most important flavonoids present in sorghum grains are 3-deoxy-anthocyanins, flavones and flavonones, and proanthocyanidins (condensed tannins) (Girard and Awika, 2018). Figure 10.1 show the several molecules that can be found in sorghum.

10.2.5 FLAVONOIDS

Flavonoids are a class of polyphenols that confers color to plant tissues. The backbone of their molecular structure consists of C_6–C_3–C_6, which are two phenyl rings and a heterocyclic ring (Aguilera-Carbó et al., 2008). Their name comes from the Latin word "Flavus," which means yellow, and, as previously mentioned, they are responsible for the color of certain fruits and plant tissues. Aside from their antioxidant and/or antimicrobial activities, there are reports that mention that flavonoids help against heart diseases, cancers, diabetes, cardiovascular or neurodegenerative diseases, among many others (Nagula and Wairkar, 2019). Sorghum is reported to contain the said molecules and it is the cereal with the highest content of this type of molecules, such as luteolin, apigenin, naringenin, quercetin, and rutin (Girard and Awika, 2018; Luo et al., 2018)

10.2.6 PHENOLIC ACIDS

As derivates of cinnamic and hydroxybenzoic acids, the skeleton of these molecules is C_6–C_1 and C_6–C_3 for its basic conformation. Phenolic acids

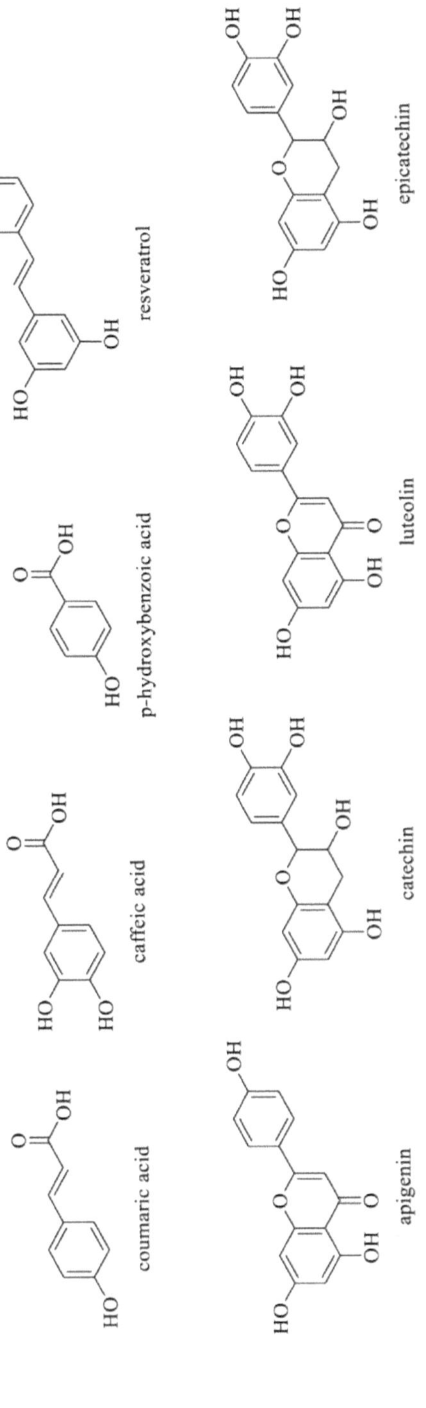

FIGURE 10.1 Polyphenolic biomolecules that can be found in sorghum.

are produced by the shikimic acid metabolic pathway and are grouped in two classes, one are the derivates of hydroxybenzoic acid and the other are from hydroxycinnamic acid (Mark et al., 2019). The most common phenolic acids are gallic acid and ellagic acid, but also caffeic, chlorogenic, cinnamic, hydroxycinnamic, among others (Aguilera-Carbó et al., 2008; Chávez-González et al., 2017). Molecules of this nature can be extracted and recovered from fruits, vegetables, tea, food waste, and biomass, as well as processed products such as wine and fruit-derived foods. In plant tissues, they can be found in free form, glycosylated, or conjugated with another polyphenols. Phenolics such as gallic, vanillic, and ferullic acids are known as hydrolyzable tannins, and they have been studied for their many bioactivities exhibited as other similar molecules, which can be found in sorghum bound to polysaccharides (Ambigaipalan et al., 2016; Miafo, 2019).

10.2.7 STILBENES

Molecules such as this possess a C_6–C_2–C_6 skeleton of two phenyl groups linked by an ethene. One of the more known stilbenes is the resveratrol, which can be found in grape skins, medicinal plants, peanuts, and berries. Other stilbenes of importance are 3′-hydroxypterostilene and pterostilbene (Jeandet et al., 2018). These compounds have been studied extensively, especially resveratrol, because of their antioxidant, anticarcinogenic, anti-inflammatory, and antidiabetic activities (Huang et al., 2019). As previously stated, one of the most studied molecule is resveratrol. Even though the resveratrol concentration in sorghum is low compared to that in the other sources, the beneficial effects are still present in human health if consumed in small amounts. It has been found in red sorghum (Bröhan et al., 2011).

Another functional activity of polyphenols, besides previously mentioned bioactivities, is the impact on gut health. When these molecules reach the large intestine, they are broken down to smaller, absorbable compounds and fermented by the microbiota and, in conjunction with some probiotics such as fructooligosaccharides, increase probiotic populations, such as *Bifidobacterium* and especially *Lactobacillus* (Ashley et al., 2019). Besides their impact on gut health, there are several studies that describe sorghum polyphenols as tumor suppressors (Vanamala et al., 2018).

10.2.8 STARCH

Starch is composed of two polymer units of D-glucose called α-amylose and amylopectin. Amylose is linked by α-1→4 glycosidic bonds and amylopectin by both α-1→4 and α-1→6 glycosidic bonds. Amylose is a linear polymer of about 1000 glucose units, while amylopectin has branches linked by α-1→6 glycosidic bonds. The physicochemical properties of starch depend largely on the ratio of amylose and amylopectin, which significantly affect the final ethanol yield (Prasad et al., 2019).

10.2.9 LIGNOCELLULOSE

The main polysaccharides present in the lignocellulosic biomass are cellulose (38%–50%), hemicellulose (23%–32%), and lignin (12%–25%). Cellulose is the most abundant polysaccharide on the earth embedded in the cell wall, along with hemicellulose and lignin polymer (Pandiyan et al., 2019). Cellulose is a long chain of D-glucose formed by glycosidic β-1→4 linkages, converting cellulose into the amorphous state due to the strong hydrogen bonding between the microfibrils. Hemicellulose (glucomannan, mannan, xylan, and xyloglucan) is distributed heterogeneously in the cell wall of the plants, together with cellulose through hydrogen bonding and lignin glycosidic bonds and ester (Liu et al., 2019). Lignin is a complex organic hydrophobic polymer, composed of phenylpropane monomers. The rigid structure of lignin strengthens the physical properties of the cell wall but makes it difficult to hydrolyze the cellulosic biomass (Ponnusamy et al., 2019).

10.3 EXTRACTION METHODOLOGIES

For bioactive compound extraction, there are several methodologies, ranging from reflux solvent extraction to emerging technologies. For recent studies, technologies such as electric pulses, ultrasound, pressurized water, and microwave have been applied to obtain bioactive compounds from several plant materials (Luo et al., 2018).

10.3.1 MICROWAVE

This is a method that serves for the extraction of soluble compounds into a fluid; water or nontoxic solvents or blends such as ethanol/water can be used. This technology has been applied to several plant materials and has been accepted as an effective alternative for bioactive compound extraction. For this method, an electromagnetic field is employed, which is generated in a frequency range of 300 MHz to 300 GHz that heats the solvent in the solid matrix, at which the solute diffuses to the surface. These fields have been used in solid plant materials that improve the extraction of phenolic bioactive compounds including flavonoids and hydroxycinnamates (Angiolillo et al., 2015). In sorghum, its dried husks were used for the extraction of bioactive compounds such as aglycones of luteolin and apigenin. Blended solvents ethanol/water and a small amount of HCl were used and showed a higher extraction yield than the conventional extraction process. These compounds are considered dyes, which can be used for antioxidant, antimicrobial, and ultraviolet protection in several products. These phytochemicals also have several beneficial effects on human health (Wizi et al., 2018).

10.3.2 ULTRASOUND EXTRACTION OF BIOACTIVE COMPOUNDS

This technology uses a frequency that exceeds 20 kHz, which is the limit for human detection for sound or hearing. The source of the ultrasound is commonly a vibrating body, which makes the medium to vibrate, and energy transfer between particles occurs from the ultrasound waves. The most effective ultrasound wave frequency ranges from 20 to 50 kHz, where it can produce cavitation, vibration, crush, or mix, and has successfully been applied for the bioactive compound extraction (Wen et al., 2018). In the case of sorghum, bran from red wholegrain was obtained by using a dehuller for decortication. Ethanol at 53% was used with a time of 21 min of ultrasonic treatment to release gallic acid equivalents, which were measured by the Folin–Ciocalteu reagent method. By HPLC–ESI–MS/MS (high-performance liquid chromatography-electrospray ionization-mass spectroscopy), procyanidins, epicatechin, taxifolin, and its hexosides were detected, which also have positive effects and applications in many industrial areas (Luo et al., 2018).

10.3.3 ELECTRIC PULSES

This technology is a nonheating processing technique that applies high-intensity short duration pulses in an electric field generated between two electrodes. This electrical procedure causes breakdown in the cell membrane by electroporation. This effect releases the intracellular contents because the permeability of the cell has been modified. In recent advances, it is suggested that pulsed electric fields could have an important effect on macromolecular interaction and their microstructure (Lohani and Muthukumarappan, 2016; Sitzmann et al., 2016). The energy that courses along the organic tissues can cause functional group ionization in molecules (carboxylic acids, amino groups) and also disrupt electrostatic interactions, resulting in the release of monomeric units from the macromolecules (Giteru et al., 2016). In sorghum, pulsed electric fields have been used on its flour. Commonly, flour is fermented to achieve or elaborate products in an easier way and to improve the bioavailability of polyphenols in flour. The disadvantage is that phenol oxidases, during fermentation, degrade polyphenolic structures and that the acidification of the media causes a rearrangement in their structures. In this case, the total phenolic content was enhanced and also higher concentrations of phenolic acids were achieved, such as protocatechuic, hydroxybenzoic, caffeic, chlorogenic, salicylic, ferulic, and chlorogenic acids (Lohani and Muthukumarappan, 2016).

10.3.4 EXTRUSION

This method consists of mechanical and thermal treatment at high pressures, low moisture levels, and low temperatures and shearing for a short time. This process is used to alter the structure of materials and in the food industry is commonly used to texturize snacks or soybean. It is versatile and has low energy demand. While extruding sorghum grain, the plant cells are being disrupted or "broken," which lowers its degree of polymerization in its structure, which includes cellulose starch and polyphenols. In a study by Morais et al. (2015), they observed that extrusion, under their conditions, removed certain types of polyphenols almost in their entirety or others decreased considerably. However, low-molecular-weight proanthocyanidins accumulated. However, if protein concentration is desired, extrusion can facilitate enzyme degradation of starch and helps accumulate sorghum protein, which eases the degradability of the grain components (Mesa-Stonestreet et al., 2015).

Several molecules have been reported as important for human health (Nash et al., 2018). As previously stated, the alternative to produce ethanol from different sources of sugars (corn, potato, and rice) should be considered for later use and consumption. Depending on the processing, certain types of compounds can be extracted by the bulk from biomass or of specific molecules are desired, the extraction must be directed in favor of the desired result (King, 2019). Polyphenol-rich fruits and residues or biomass can be considered as a mixture for compound accumulation by ethanolic extraction. According to previous statements, compounds such as resveratrol, coumaric acid, and flavonoids are commonly detected with the use of a chromatograph, such as an HPLC with a C-18 column, a photodiode array detector, and a UV–Visible spectrophotometry detector. For more efficient identification results, if able, a mass spectroscopy analysis can be added (Ascacio-Valdés et al., 2016). If the phenolic content analysis is required, a simpler spectrophotometric technique is often used, which is the Folin–Ciocalteu method where the phenolics, in presence of the reagent, can be quantified at 750 nm (Wong-Paz et al., 2015).

10.4 BIOTECHNOLOGICAL PROCESSING

The main aspect that can be taken into consideration for microbial growth during the processing and extraction of metabolites is the sugar content in sorghum grain, juice, and lignocellulose. Bioprocessing is useful for several reasons regarding sorghum. In Africa, sorghum is consumed in a presentation of porridge-like fermented food, which improves protein digestibility. Also, red sorghum flour has been fermented with Lactobacilli for flatbread production. Its flavor was unaffected and flavonoid content was increased. For processing starch or flour, enzymes have been used to decrease the polymerization degree and release bioactive phenolic compounds. This must be related to the compounds being bonded to polysaccharides (Miafo et al., 2019), and the presence of enzymes, such as amylases, glucosidases, among others, helps to release bioactive compounds into the flour. The polysaccharide content of sorghum can also be used in biotechnological terms from a microbial platform or biofactory point of view. Several microorganisms have been proposed in this context for the wide range of useful chemicals that they can produce. These are known for the several products of their metabolism, which are from solventogenesis, organic acid production, cellulose disruption, or derivates of polysaccharides, and later monomer degradation. The advantage of most of these traits is that its production can initiate from sugars, which may come from renewable sources, such as starch or lignocellulose.

10.4.1 ABE FERMENTATION

From the ABE fermentation, acetone, butanol, and ethanol can be produced in a 3:6:1 ratio, which puts butanol in the front because of its higher yields. From these solvents, ethanol is considered as a biofuel for car engines, as well as butanol and acetone as a solvent itself or as a precursor of other compounds (Oliva-Rodríguez et al., 2019). For the ABE fermentation process, the mostly used bacteria are *Clostridium acetobutylicum* and *Clostridium beijerinckii*, where several sources of sugar have been used, which include sorghum. Another advantage, concerning starch, is that Clostridia is able to produce the enzymes required to degrade starch, accumulate glucose, and produce solvents. In the case of sugars from the lignocellulosic origin, cellulose and hemicellulose must be previously degraded for their sugars to be metabolized to solvents (Qin et al., 2018). Pentoses such as xylose and arabinose, which are found commonly in hemicellulose hydrolysates, are sugars that can be consumed by solventogenic bacteria but with less efficiency compared to glucose. There are reports where sorghum has been used for ABE fermentation processes from both starch and its lignocellulose (Cai et al., 2013; Mirfakhar et al., 2017). There is an application of ABE fermentation that has an impact on the food industry that involves butanol, which is the product of butyl butyrate. This molecule has applications in biofuels, beverages and food. It is a colorless liquid that has a fruity odor. It involves the butyric acid and butanol production in ABE fermentation and the addition of lipases to bind the molecules in an ester bond to form the compound of interest (Xin et al., 2019).

10.4.2 LACTIC ACID FERMENTATION AND SHORT-CHAIN FATTY ACID PRODUCTION

This compound has extensive applications in food, textile, pharmaceutical, and many other areas. Commonly, Lactobacilli are used for lactic acid production (Pranoto et al., 2013). From the available polysaccharides in sorghum (stalk juice, starch, and lignocellulose), fermentations are viable for lactic acid production. There is a study where a facultative anaerobic strain of *Lactobacillus* sp. is used. From the juice of sweet sorghum stalks, lactic acid has been produced. Formerly, it was produced by chemical synthesis, but by biotechnological means, the acid is produced with fewer resources (Hetényiet al., 2010). Lactobacilli are bacteria that can produce fatty acids from sugars, such as propionic, acetic, and butyric acids. The latter compounds are fatty acids of short chains, which have applications in the food industry. In this case, the sugars that can be obtained from starch or lignocellulose may represent a source of these compounds by bioprocessing. For instance, there is a report by Veeravalli and Mathews (2018), where they were able to produce lactic acid and acetic acid from xylose by fermentation with *Lactobacillus buchnerii*. Xylose in hemicellulose is one of its major sugars and is obtained from the pretreatment of the heteropolysaccharide by enzymes or chemical treatment. Studies have been published related to the lignocellulosic biomass from sorghum to obtain fermentable sugars, and a fraction of them is xylose, which can be used as previously stated (Jagtap and Rao, 2018).

10.4.3 XYLITOL

This molecule is a polyalcohol or a pentahydroxylated molecule derived from xylose. It has been applied as a sweetener in the food industry, focusing on diabetic people because xylitol is metabolized in a different pathway than sugars. Also, it has been used as an anticarcinogenic in pharmaceutical industries. As its production by the chemical way is costly, because of requirements of high temperatures, high pressures, and Ni as a catalyst, microbial production is more feasible. By using hemicellulose hydrolysates from sorghum, xylitol production was achieved by xylose fermentation by *D. hansenii* (Ledezma-Orozco et al., 2018). For its production, a process similar to bioethanol production can be carried out, from pretreating the material and hydrolyzing hemicellulose to yield xylose, where strains such as *D. hansenii* or *C. guillermondii* produce xylitol (Venkateswar et al., 2015).

10.4.4 ETHANOL FROM STARCH

The production of ethanol from grain sorghum (starch) consists of the preparation of a mixture of sorghum flour with water, and then it is liquefied by thermostable α-amylase to decompose the starch molecules into dextrins and other smaller molecules. Subsequently, the liquefied slurry is saccharified to convert dextrins into glucose using a glucoamylase. Finally, the enzymatic hydrolysate obtained is subjected to fermentation (using mainly yeasts) for the production of ethanol, which is recovered by distillation (Figure 10.2) (Khlestkin et al., 2018). The conventional process of producing ethanol from lignocellulosic materials is summarized in three steps: (i) physical, chemical, or biological pretreatment of biomass; (ii) enzymatic hydrolysis (saccharification); and (iii) fermentation (Gottumukkala et al., 2017).

10.4.5 ETHANOL FROM STALK JUICE

Sweet sorghum juice was obtained by mechanically triturating the stems harvested using roll mills or screw press (Appiah-Nkansahet al., 2018). The juice obtained is purified with calcium hydroxide $Ca(OH)_2$ or calcium saccharate to a maximum concentration of 2% (w/w) according to the concentration of solids in the juice. The addition of these compounds increases the pH, reduces the dyes present, neutralizes the content of organic acids, and generates calcium phosphate precipitates, which can be removed by sedimentation (Zabed et al., 2017). Afterward, the juice is filtered and evaporated to a concentration of 14%–18% (v/v), depending on the tolerance of the fermenting microorganisms. The concentrated juice is supplemented with ammonium sulfate or other nitrogen sources to be finally subjected to fermentation (preferably yeast) to produce ethanol and recovered through a distillation process (Figure 10.3) (Appiah-Nkansahet al., 2019).

10.5 CONCLUDING REMARKS

There is great potential in exploiting sorghum. It is an important food source in Africa and Asia by their own processing recipes to produce several food types, and its consumption carries several health benefits.

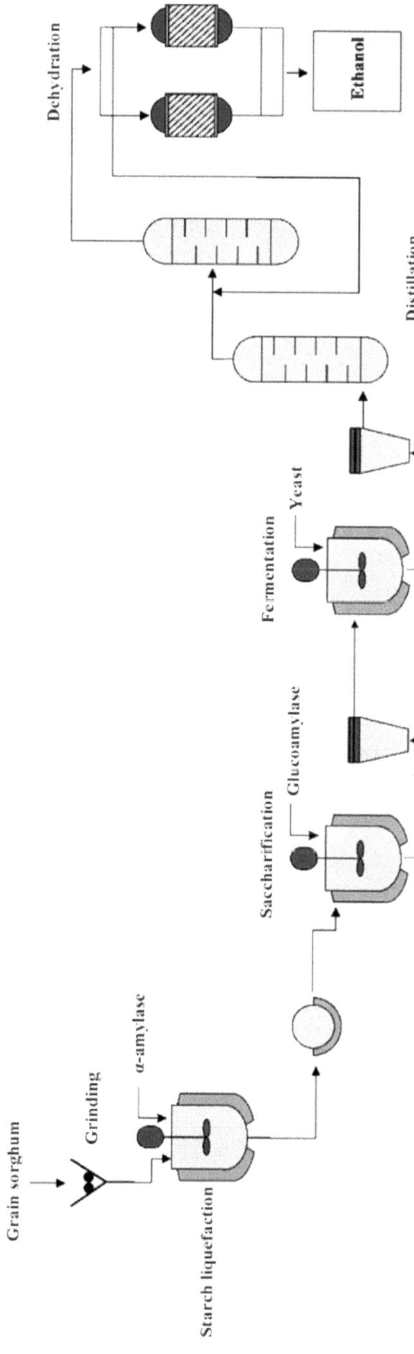

FIGURE 10.2 Schematic diagram of ethanol production from sorghum grains.

By extracting or making more bioavailable the wide array of useful biomolecules, enhancement of sorghum takes place while also taking into consideration the fact that it is a very versatile crop from an agricultural point of view. The presence of phytochemicals of interest, as well as the polysaccharide content, makes sorghum a bioresource for biotechnological processing, which used starch, lignocellulose, or stalk juice to produce another wide array of high added-value compounds, which paired with native biomolecules puts sorghum into spotlight among other crops.

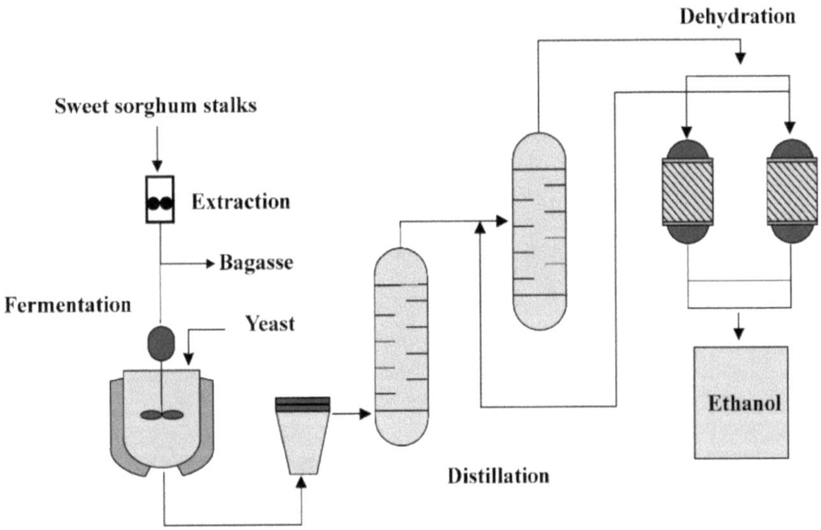

FIGURE 10.3 Schematic diagram of ethanol production from sorghum stalk juice.

KEYWORDS

- sorghum
- polyphenols
- polysaccharides
- food technology
- biotechnology

REFERENCES

Aguilera-Carbo, A.; Augur, C.; Prado-Barragan, L.A.; Favela-Torres, E.; Aguilar, C.N. Microbial production of ellagic acid and biodegradation of ellagitannins. *Appl. Microbiol. Biotechnol.* 2008, 78, 189–199.

Ahmad Dar, R.; Ahmad Dar, E.; Kaur, A.; Gupta Phutela, U. Sweet sorghum-a promising alternative feedstock for biofuel production. *Renew. Sustain. Energy Rev.* 2018, 82, 4070–4090.

Amaducci, A.; Monti, A., Venturi, G. Non-structural carbohydrates and fiber components in sweet and fiber sorghum as affected by low and normal input techniques. *Ind. Crops Prod.* 2004, 20, 111–118.

Ambigaipalan, P.; Costa De Camargo, A.; Shahidi, F.; Priyatharini-Ambigaipalan, M.N.; De Camargo, A.C. Identification of phenolic antioxidants and bioactives of pomegranate seeds following juice extraction using HPLC-DAD-ESI-MS. *Food Chem.* 2016, 221, 1883–1894.

Angiolillo, L.; Del Nobile, M.A.; Conte, A. The extraction of bioactive compounds from food residues using microwaves. *Curr. Opin. Food Sci.* 2015, 5, 93–98.

Appiah-Nkansah, N.B.; Li, J.; Rooney, W.; Wang, D. A review of sweet sorghum as a viable renewable bioenergy crop and its techno-economic analysis. *Renew. Energy* 2019, 143, 1121–1132.

Appiah-Nkansah, N.B.; Zhang, K.; Rooney, W.; Wang, D. Ethanol production from mixtures of sweet sorghum juice and sorghum starch using very high gravity fermentation with urea supplementation. *Ind. Crops Prod.* 2018, 111, 247–253.

Arenas-Cárdenas, P.; López-López, A.; Moeller-Chávez, G.E.; León-Becerril, E. Current pretreatments of lignocellulosic residues in the production of bioethanol. *Waste Biomass Valor.* 8, 2016, 161–181.

Aruna, C.; Visarada, K.B.R.S. Other industrial uses of sorghum. In *Breeding Sorghum for Diverse End Uses*; Aruna, C.; Visarada, K.B.R.S.; Bhat, B.; Tonapi, V. (Eds.); Elsevier: UK, 2019; p. 271.

Ascacio-Valdés, J.A.; Aguilera-Carbó, A.F.; Buenrostro, J.; Prado-Barragán, A.; Rodríguez-Herrera, R.; Aguilar, C.N. The complete biodegradation pathway of ellagitannins by *Aspergillus niger* in solid-state fermentation. *J. Basic Microbiol.* 2016, 56, 329–336.

Ashley, D.; Marasini, D.; Brownmiller, C.; Ae-Lee, J., Carbonero, F.; Lee, S.O. Impact of grain sorghum polyphenols on microbiota of normal weight and overweight/obese subjects during *in vitro* fecal fermentation. *Nutrients* 2019, 11, 217.

Birt, D.F.; Boylston, T.; Hendrich, S.; Jane, J.L.; Hollis, J.; Li, L.; McClelland, J.; Moore, S.; Phullips, G.; Rowling, M.; Schalinske, K.; Scott, M.P; Whitley, E. Resistant starch: Promise for improving human health. *Adv. Nutr.* 2013, 4, 587–601.

Bröhan, M.; Jerkovic, V.; Collin, S. Potentiality of red sorghum for producing stilbenoid-enriched beers with high antioxidant activity. *J. Agric. Food Chem.* 2011, 59, 4088–4094.

Buenrostro-Figueroa, J.; Ascacio-Valdés, A.; Sepúlveda, L.; De La Cruz, R.; Prado-Barragán, A.; Aguilar-González, M.A.; Aguilar, C.N. Potential use of different agroindustrial by-products as supports for fungal ellagitannase production under solid-state fermentation. *Food Bioprod. Process.* 2013, 92, 376–382.

Cai, D.; Zhang, T.; Zheng, J.; Chang, Z.; Wang, Z.; Qin, P.; Yong, T.; Wei, T. Biobutanol from sweet sorghum bagasse hydrolysate by a hybrid pervaporation process. *Bioresour. Technol.* 2013, 145, 97–102.

Chávez-González, M.L.; Guyot, S.; Rodríguez-Herrera, R.; Prado-Barragán, A.; Aguilar, C.N. Exploring the degradation of gallotannins catalyzed by tannase produced by *Aspergillus niger* GH1 for ellagic acid production in submerged and solid-state fermentation. *Appl. Biochem. Biotechnol.* 2017, 185, 476–483.

Dar, R.A.; Dar, E.A.; Kaur, A.; Phutela, U.G. Sweet sorghum—a promising alternative feedstock for biofuel production. *Renew. Sustain. Energy Rev.* 2018, 82, 4070–4090.

Dey, T.B.; Kuhad, R.C. Enhanced production and extraction of phenolic compounds from wheat by solid-state fermentation with *Rhizopus oryzae* RCK2012. *Biotechnol. Rep.* 2014, 4, 120–127.

do Prado, A.C.P.; da Silva, H.S.; da Silveira, S.M.; Barreto, P.L.M.; Vieira, C.R.W.; Maraschin, M.; Ferreira, S.R.S.; Block, J.M. Effect of the extraction process on the phenolic compounds profile and the antioxidant and antimicrobial activity of extracts of pecan nut [*Caryaillinoinensis* (Wangenh) C. Koch] shell. *Ind. Crops Prod.* 2014, 52, 552–561.

Fadel, A.; Mahmoud, A.M.; Ashworth, J.J.; Li, W.; Ng, Y.L.; Plunkett, A. Health-related effects and improving extractability of cereal arabinoxylans. *Int. J. Biol. Macromol.* 2018, 109, 819–831.

Garaguso, I.; Nardini, M. Polyphenols content, phenolics profile and antioxidant activity of organic red wines produced without sulfur dioxide/sulfites addition in comparison to conventional red wines. *Food Chem.* 2015, 179, 336–342.

Girard, A.L.; Awika, J.M. Sorghum polyphenols and other bioactive components as functional and health promoting food ingredients. *J. Cereal Sci.* 2018, 84, 112–124.

Giteru, S.G.; Oey, I.; Ali, M.A. Feasibility of using pulsed electric fields to modify biomacromolecules: A review. *Trends Food Sci. Technol.* 2018, 72, 91–113.

Gomes, L.; Almeida, F.; Augusto, R.; Lúcia, M.; Simeone, F.; César, P.; Ribeiro, D.O.; Soares, A.; Sylvio, A.; Gonçalves, A.; Eugene, R. Composition and growth of sorghum biomass genotypes for ethanol production. *Biomass Bioenergy* 2019, 122, 343–348.

Gottumukkala, L.D.; Haigh, K.; Görgens, J. Trends and advances in conversion of lignocellulosic biomass to biobutanol: Microbes, bioprocesses and industrial viability. *Renew. Sustain. Energy Rev.* 2017, 76, 963–973.

Hetényi, K.; Gál, K.; Németh, Á.; Sevella, B. Use of sweet sorghum juice for lactic acid fermentation: Preliminary steps in a process optimization. *J. Chem. Technol. Biotechnol.* 2010, 85, 872–877.

Huang, X.; Li, X.; Xie, M.; Huang, Z.; Huang, Y.; Wu, G.; Peng, Z.; Sun, Y.; Ming, Q.; Liu, Y.; Chen, J.; Xu, S. Resveratrol: Review on its discovery, anti-leukemia effects and pharmacokinetics. *Chem. Biol. Interact.* 2019, 306, 29–38.

Impa, S.M.; Perumal, R.; Bean, S.R.; John Sunoj, V.S.; Jagadish, S.V.K.Water deficit and heat stress induced alterations in grain physico-chemical characteristics and micronutrient composition in field grown grain sorghum. *J. Cereal Sci.* 2019, 86, 124–131.

Jagtap, S.S.; Rao, C.V. Microbial conversion of xylose into useful bioproducts. *Appl. Microbiol. Biotechnol.* 2018, 102, 9015–9036.

Jeandet, P.; Sobarzo-Sánchez, E.; Clément, C.; Nabavi, S.F.; Habtemariam, S.; Mohammad, S.: Cordelier, S. Engineering stilbene metabolic pathways in microbial cells. *Biotechnol. Adv.* 2018, 36, 2264–2283.

Kaur, Y.R.; Uppal, S.K.; Sharma, P.; Oberoi, H.S. Chemical composition of sweet sorghum juice and its comparative potential of different fermentation processes for enhanced ethanol production. *Sugar Tech* 2010, 15, 305–310.

Khlestkin, V.K.; Peltek, S.E.; Kolchanov, N.A. Review of direct chemical and biochemical transformations of starch. *Carbohydr. Polym.* 2018, 181, 460–476.

King, J.W. The relationship between cannabis/hemp use in foods and processing methodology. *Curr. Opin. Food Sci.* 2019, 28, 32–40.

Kulamarva, A.G.; Sosle, V.R.; Raghaven, G.S.V. Nutritional and rheological properties of sorghum. *Int. J. Food Prop.* 2009, 12, 55–69.

Kumar, V.; Sharma, A.; Kaur, S.; Bali, S.; Sharma, M.; Kumar, R.; Bhardwaj, R.; Kumar, A.; Thukrai, A. Differential distribution of polyphenols in plants using multivariate techniques. *Biotechnol. Res. Innov.* 2019, 3, 1–21.

Léder, I. Sorghum and millets, in cultivated plants, primarily as food sources. In *Encyclopedia of Life Support Systems (EOLSS), Developed under the Auspices of the UNESCO*; György F. (Ed.); Eolss Publishers: Oxford, UK, 2004; pp. 1–18

Ledezma-Orozco, E.; Ruíz-Salazar, R.; Bustos-Vázquez, G.; Montes-García, N.; Roa-Cordero, V.; Rodríguez-Castillejos, G. Producción de xilitol a partir de hidrolizados ácidos no detoxificados de bagazo de sorgo por. *Agrociencia* 2018, 52, 1095–1106.

Lima, M.C.; Paiva de Sousa, C.; Fernández-Prada, C.; Harel, J.; Dubreuil, J. D.; de Souza, E. L. A review of the current evidence of fruit phenolic compounds as potential antimicrobials against pathogenic bacteria. *Microb. Pathog.* 2019, 130, 259–270.

Liu, G.; Gilding, E.K.; Kerr, E.D.; Schulz, B.L.; Tabet, B.; Hamaker, B.R.; Godwin, I.D. Increasing protein content and digestibility in sorghum grain with a synthetic biology approach. *J. Cereal Sci.* 2019, 85, 27–34.

Lohani, U.C.; Muthukumarappan, K. Application of the pulsed electric field to release bound phenolics in sorghum flour and apple pomace. *Innov. Food Sci. Emerg. Technol.* 2016, 35, 29–35.

Lolasi, F.; Amiri, H.; Asadollahi, M.; Karimi, K. Using sweet sorghum bagasse for production of amylases required for its grain hydrolysis via biorefinery platform. *Ind. Crops Prod.* 2018, 128, 473–481.

López-Trujillo, J.; Medina-Morales, M.A.; Sánchez-Flores, A.; Arévalo, C.; Ascacio-Valdés, J.A.; Mellado, M.; Aguilar, C.N.; Aguilera-Carbó, A.F. Solid bioprocess of tarbush (*Flourensia cernua*) leaves for β-glucosidase production by *Aspergillus niger*: Initial approach to fiber–glycoside interaction for enzyme induction. *3 Biotech* 2017, 7, 271.

Luo, X.; Cui, J.; Zhang, H.; Duan, Y.; Zhang, D.; Cai, M.; Chen, G. Ultrasound assisted extraction of polyphenolic compounds from red sorghum (*Sorghum bicolor* L.) bran and their biological activities and polyphenolic compositions. *Ind. Crops Prod.* 2018, 112, 296–304.

Majdalawieh, A.F.; Mansour, Z.R. Sesamol, a major lignan in sesame seeds (*Sesamum indicum*): Anti-cancer properties and mechanisms of action. *Eur. J. Pharmacol.*, 2019, 855, 75–89.

Makris, D. P. Green extraction processes for the efficient recovery of bioactive polyphenols from wine industry solid wastes—Recent progress. *Curr. Opin. Green Sustain. Chem.* 2018, 13, 50–55.

Mark, R.; Lyu, X.; Lee, J.J.L.; Parra-Saldívar, R.; Chen, W.N. Sustainable production of natural phenolics for functional food applications. *J. Funct. Foods* 2019, 57, 233–254.

Medina, M.A; Belmares, R.E.; Aguilera-Carbo, A.; Rodríguez-Herrera, R.; Aguilar, C.N. Fungal culture systems for production of antioxidant phenolics using pecan nut shells as sole carbon source. *Am. J. Agric. Biol. Sci.* 2010, 5, 397–402.

Mesa-Stonestreet, N.J.; Alavi, S.; Gwirtz, J. Extrusion-enzyme liquefaction as a method for producing sorghum protein concentrates. *J. Food Eng.* 2012, 108, 365–375.

Miafo, A.P.T.; Koubala, B.B.; Kansci, G.; Muralikrishna, G. Free sugars and non-starch polysaccharides–phenolic acid complexes from bran, spent grain and sorghum seeds. *J. Cereal Sci.* 2019, 87, 124–131.

Mirfakhar, M.; Asadollahi, M.A.; Amiri, H.; Karimi, K. Phenolic compounds removal from sweet sorghum grain for efficient biobutanol production without nutrient supplementation. *Ind. Crops Prod.* 2017, 108, 225–231.

Morais, L.; Silva, S.; Carvalho, P.; Vieira, V.; Menezes, C.B.; Bandeira, A.; Ribeiro, A.; Awika, J.M.; Martino, H.; Maria, H.; Pinheiro-Sant'Ana, H. Phenolic compounds profile in sorghum processed by extrusion cooking and dry heat in a conventional oven. *J. Cereal Sci.* 2015, 65, 220–226.

Nagula, R. L.; Wairkar, S. Recent advances in topical delivery of flavonoids: A review. *J. Control. Release* 2019, 296, 190–201.

Nash, V.; Ranadheera, C.S.; Georgousopoulou, E.N.; Mellor, D.D.; Panagiotakos, D.; McKune, A.; Kellett, J.; Naumovski, N. The effects of grape and red wine polyphenols on gut microbiota—A systematic review. *Food Res. Int.* 2018, 113, 277–287.

Oliva-Rodríguez, A.G.; Quintero, J.; Medina-Morales, M.A.; Morales-Martínez, T.K.; Rodríguez-de la Garza, J.A.; Moreno-Dávila, M.; Aroca, G.; Ríos-González, L.J. *Clostridium* strain selection for co-culture with Bacillus subtilis for butanol production from agave hydrolysates. *Bioresour. Technol.* 2019, 275, 410–415.

Pandiyan, K.; Singh, A.; Singh, S.; Saxena, A.K.; Nain, L. Technological interventions for utilization of crop residues and weedy biomass for second generation bio-ethanol production. *Renew. Energy* 2019, 132, 723–741.

Ponnusamy, V.K.; Nguyen, D.D.; Dharmaraja, J.; Shobana, S.; Banu, J.R.; Saratale, R.G.; Chang, S.W.; Kumar, G. A review on lignin structure, pretreatments, fermentation reactions and biorefinery potential. *Bioresour. Technol.* 2019, 271, 462–472.

Pranoto, Y.; Anggrahini, S.; Efendi, Z. Effect of natural and *Lactobacillus plantarum* fermentation on in-vitro protein and starch digestibilities of sorghum flour. *Food Biosci.* 2013, 2, 46–52.

Prasad, R.K.; Chatterjee, S.; Mazumder, P.B.; Gupta, S.K.; Sharma, S.; Vairale, M.G.; Datta, S.; Dwivedi, S.K.; Gupta, D.K. Bioethanol production from waste lignocelluloses: A review on microbial degradation potential. *Chemosphere* 2019, 231, 588–606

Qin, Z.; Duns, G.J.; Pan, T.; Xin, F. Consolidated processing of biobutanol production from food wastes by solventogenic *Clostridium* sp. strain HN4. *Bioresour. Technol.* 2018, 264, 148–153.

Ramatoulaye, F.; Mady, C.; Fallou, S.; Amadou, K.; Cyril, D.; Massamba, D. Production and use sorghum: A literature review. *J. Nutr. Health Food Sci.* 2016, 4, 1–4.

Rojas, R.; Álvarez-Pérez, O.B.; Contreras-Esquivel, J.C.; Vicente, A.; Flores, A.; Sandoval, J.; Aguilar, C.N. Valorisation of mango peels: Extraction of pectin and antioxidant and antifungal polyphenols. *Waste Biomass Valor.* 2020, 11, 89–98.

Rodrigues de Sousa, A.; Castro Moreira, M.E.;Grancieri, M.; Lopes Toledo, R.C.; Oliveira de Araújo, F.; Cuquetto Mantovani, H.; Vieira Queiroz, V.A.; Duarte Martino, H.S.

Extruded sorghum (*Sorghum bicolor* L.) improves gut microbiota, reduces inflammation, and oxidative stress in obese rats fed a high-fat diet. *J. Funct. Foods* 2019, 58, 282–291.

Samarth, A.G.; More, D.R.; Hashmi, I. Studies on physico-chemical properties and nutritional profile of sweet sorghum. *Int. J. Chem. Stud.* 2018, 6, 2826–2828.

Shen, R.L.; Zhang, W.L.; Dong, J.L.; Ren, G.X.; Chen, M. Sorghum resistant starch reduces adiposity in high-fat diet-induced overweight and obese rats via mechanisms involving adipokines and intestinal flora. *Food Agric. Immunol.* 2015, 26, 120–130.

Sepúlveda, L.; De la Cruz, R.; Buenrostro, J.; Ascacio-Valdés, J.A.; Aguilera-Carbó, A.F.; Prado, A.; Rodríguez-Herrera, R.; Aguilar, C.N. Effect of different polyphenol sources on the efficiency of ellagic acid release by *Aspergillus niger*. *Rev. Argent. Microbiol.* 2016, 48, 71–77.

Stefoska-Needham, A.; Beck, E.J.; Johnson, S.K.; Tapsell, L.C. Sorghum: An underutilized cereal whole grain with the potential to assist in the prevention of chronic disease. *Food Rev. Int.* 2015, 31, 401–437.

Simnadis, T.G.; Tapsell, L.C.; Beck, E.J. Effect of sorghum consumption on health outcomes: a systemic review. *Nutr. Rev.* 2016, 74, 690–707.

Singh, R. B.; Gupta, A. K.; Fedacko, J.; Juneja, L.R.; Jarcuska, P.; Pella, D. Effects of diet and nutrients on epigenetic and genetic expressions. In *The Role of Functional Food Security in Global Health*; Watson, R.; Singh, R.; Takahashi, T. (Eds.); Elsevier Inc. UK, 2019; pp. 681–707.

Sitzmann, W.; Vorobiev, E.; Lebovka, N. Applications of electricity and specifically pulsed electric fields in food processing: Historical backgrounds. *Innov. Food Sci. Emerg. Technol.* 2016, 37, 302–311.

Vanamala, J.K.P.; Massey, A.R.; Pinnamaneni, S.R.; Reddivari, L.; Reardon, K.F. Grain and sweet sorghum (*Sorghum bicolor* L. Moench) serves as a novel source of bioactive compounds for human health. *Crit. Rev. Food Sci. Nutr.* 2018, 58, 2867–2881.

Veeravalli, S.S.; Mathews, A.P. Continuous fermentation of xylose to short chain fatty acids by *Lactobacillus buchneri* under low pH conditions. *Chem. Eng. J.* 2018, 337, 764–771.

Venkateswar, L.; Goli, J.K.; Gentela, J.; Koti, S. Bioconversion of lignocellulosic biomass to xylitol: An overview. *Bioresour. Technol.* 2015, 213, 299–310.

Verdugo-Valdez, A.; Segura Garcia, L.; Kirchmayr, M.; Ramírez Rodríguez, P.; González Esquinca, A.; Coria, R.; Gschaedler Mathis, A. Yeast communities associated with artisanal mezcal fermentations from *Agave salmiana*. *Antonie van Leeuwenhoek*, 2011, 100, 497–506.

Vila-Real, C.; Pimenta-Martins, A.; Ndegwa, H.; Gomes, A.M.; Pinto, E. Nutritional value of African indigenous whole grains cereals millet and sorghum. *Nutr. Food Sci. Int. J.*, 2017, 4, 1–4.

Wen, C.; Zhang, J.; Zhang, H.; Dzah, C.S.; Zandile, M. Advances in ultrasound assisted extraction of bioactive compounds from cash crops—A review. *Ultrason. Sonochem.* 2018, 48, 538–549.

Wizi, J.; Wang, L.; Hou, X.; Tao, Y.; Ma, B.; Yang, Y. Ultrasound-microwave assisted extraction of natural colorants from sorghum husk with different solvents. *Ind. Crops Prod.* 2018, 120, 203–213.

Wong-Paz, J.E.; Contreras-Esquivel, J.C.; Rodríguez-Herrera, R.; Carrillo-Inungaray, M.L.; López, L.; Nevárez-Moorillón, G.; Aguilar, C.N. Total phenolic content in vitro antioxidant activity and chemical composition of plant extracts from semiarid Mexican region. *Asian Pac. J. Trop. Med.* 2015, 8, 104–111.

Xin, F.; Zhang, W.; Jiang, M. Forum bioprocessing butanol into more valuable butyl butyrate. *Trends Biotechnol.* 2019, 37, 923–926.

Zabed, H.; Sahu, J.N.; Suely, A.; Boyce, A.N.; Faruq, G. Bioethanol production from renewable sources: Current perspectives and technological progress. *Renew. Sustain. Energy Rev.* 2017, 71, 475–501.

Zhu, F. Structure, physicochemical properties, modifications, and uses of sorghum starch. *Compr. Rev. Food Sci. F.* 2014, 13, 597–610.

CHAPTER 11

EFFICIENCY OF FERTILIZER APPLICATION ON SPRING WHEAT IN THE CONDITIONS OF THE URAL REGION

EKATERINA A. SEMENOVA[1], RAFAIL A. AFANAS'EV[2], and MICHAEL SMIRNOV[1*]

[1]*Agrochemical Center of the Sverdlovsk, d. 109, Furmanova Street, Ekaterinburg 620144, Russia*

[2]*Pryanishnikov All-Russian Scientific Research Institute of Agrochemistry, d. 31A, Pryanishnikova Street, Moscow 127550, Russia*

[*]*Corresponding author. E-mail: natnatali1988@gmail.com, rafail-afanasev@mail.ru*

ABSTRACT

The chapter considers the results of a three-year field experiment (2015–2017) on fertilization of spring wheat cultivated in soil and climatic conditions of the Ural region. The experiment is based on the cultivated gray forest soil under the 14-variant scheme of mineral fertilizer (N—nitrogen; P—phosphorus; K—potassium) application in rates from 30 to 120 kg/ha of each element applied against the background of the other two elements. The variety of spring wheat was Simbirtsit, and the precursor was black steam. As a result of the research studies, it is revealed the influence of nitrogen, phosphoric, and potash fertilizers on the productivity and quality of grains depending on the concrete soil and weather conditions. In particular, it is shown that with an increase in the rate of nitrogen from N30 to N120 against the background of P60K60, the grain yield increased in the conditions of drought in 2016 and in the most favorable precipitation and temperature conditions in 2015. More complex dependences in these conditions are established by the efficiency of phosphoric and potash fertilizers.

11.1 INTRODUCTION

In connection with the harsh climatic conditions of the Ural region, complicating the cultivation of grain crops, the leading grain crop here is spring wheat (Koshkin, 2010). This is largely due to the severe climatic conditions prevailing in the region, characterized by low temperatures of the winter period, which cannot withstand winter wheat (Figures 11.1–11.4). In the Sverdlovsk region, there are milder, compared with other areas of the Urals, winter temperatures; however, in this area, the acreage of winter wheat is extremely limited.

The area of arable land with gray forest soils, including light gray, gray, and dark gray, in the Ural region exceeds 3 million ha (Ministry of Agriculture of Russian Federation, 2008), so studies to optimize the mineral nutrition of plants in this region, as in other regions of the country, are important (Kidin, 2012; Mineev, 2005; Timiryazev, 2006; Pryanishnikov, 1965; Mineev et al., 2017). The features of the Trans-Urals are also soil conditions, namely, the relatively high agrochemical indicators of fertility in the pH, humus content, and mobile forms of potassium and phosphorus (Tables 11.1–11.3).

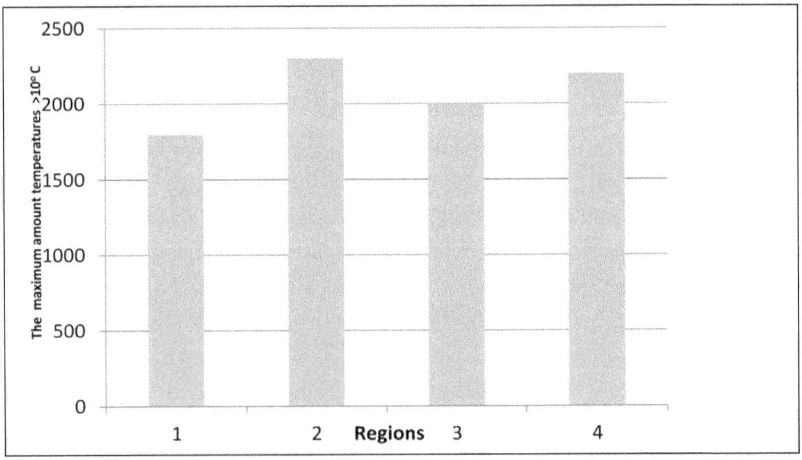

FIGURE 11.1 Maximum temperatures >10 °C in regions. 1—Sverdlovsk; 2—Chelyabinsk; 3—yumen; 4—Kurgan.

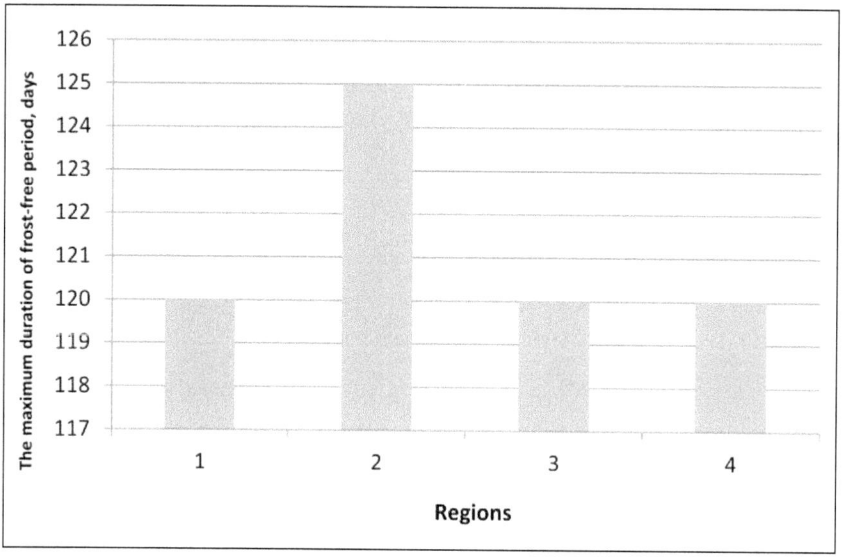

FIGURE 11.2 Maximum duration of frost-free period in regions. 1——Sverdlovsk; 2—Chelyabinsk; 3—Tyumen; 4—Kurgan.

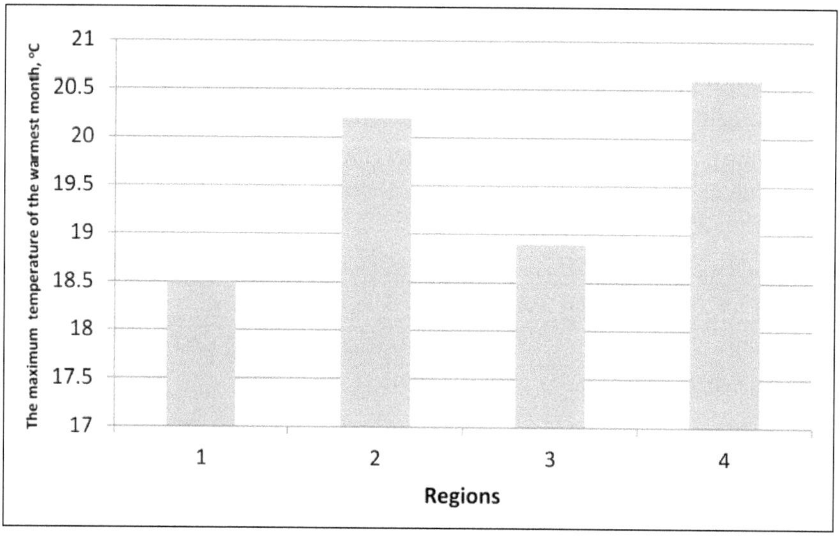

FIGURE 11.3 Maximum temperature of the warmest month (°C) in regions. 1—Sverdlovsk; 2—Chelyabinsk; 3—Tyumen; 4—Kurgan.

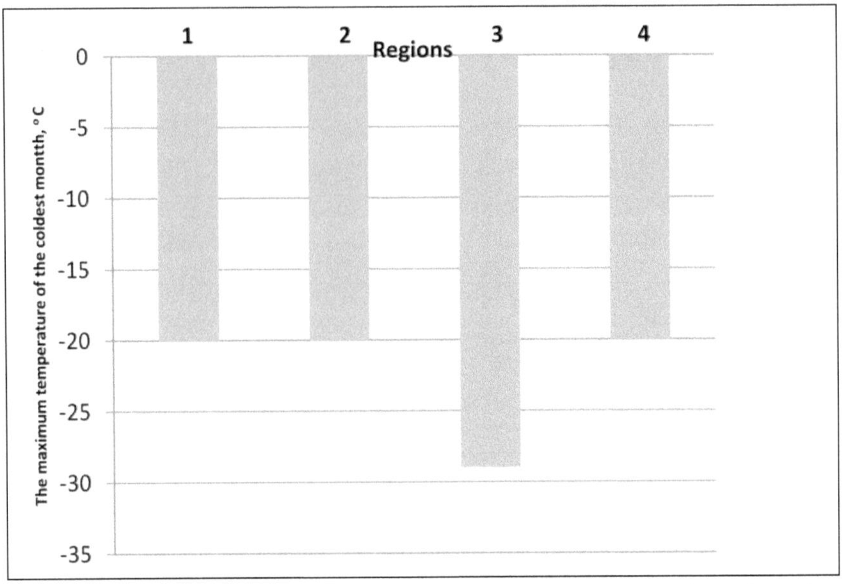

FIGURE 11.4 Maximum temperature of the coldest month (°C) in regions. 1—Sverdlovsk; 2—Chelyabinsk; 3—Tyumen; 4—Kurgan.

The area of arable land with strongly and medium acidic soils in the region does not exceed 10%, with humus content less than 4%, the area is no more than 30% of arable land, and with low to medium content of exchangeable potassium, the area is no more than 15%. At the same time, more than 50% of the arable land is characterized by a low content to very low content of mobile phosphorus, and about 30% of the arable land is characterized by an average content.

TABLE 11.1 Area of Arable Land With Acid Soils in the Ural Region

Soil Acidity	Percentage of Total Area
Highly acidic soils	1.6
Acid soils	7.8
Highly acidic soils	36.3
All acidic soils	45.6

Up to 13% of the arable wedge in the region belongs to the solonetz complexes exposed to wind and water erosion and therefore needs reclamation improvement.

TABLE 11.2 Area of Arable Land With Different Humus Contents in the Ural Region

Content of Humus in the Soil,%	Percentage of Total Area
<2	3.1
2.1–4.0	26.9
4.1–6.0	46.8
6.1–8.0	19.2
8.1–10.0	3.6
>10	0.6

TABLE 11.3 Areas of Arable Soils of the Ural Region With Different Contents of Mobile Potassium and Phosphorus

Content of Humus in the Soil	Percentage of Total Area	
	Content of Mobile Potassium	Content of Mobile Phosphorus
Very low	0.1	11.3
Low	2.2	40.3
Medium	10.5	30.4
Elevated	21.5	9
High	29.8	4.7
Very high	36.9	4.3

According to the main agroclimatic indicators, the Ural region, in general, belong to the zone of insufficient moisture (Figures 11.5–11.8).

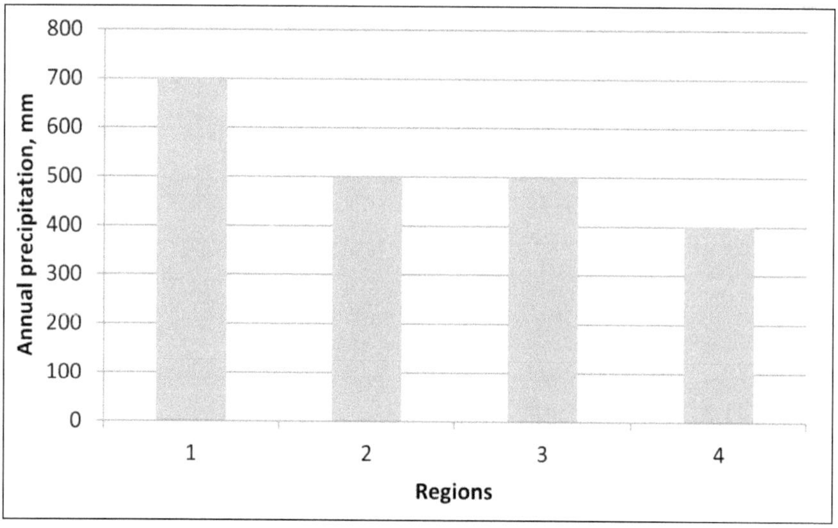

FIGURE 11.5 Annual rainfall in the regions of Ural (mm). 1—Sverdlovsk; 2—Chelyabinsk; 3—Tyumen; 4—Kurgan.

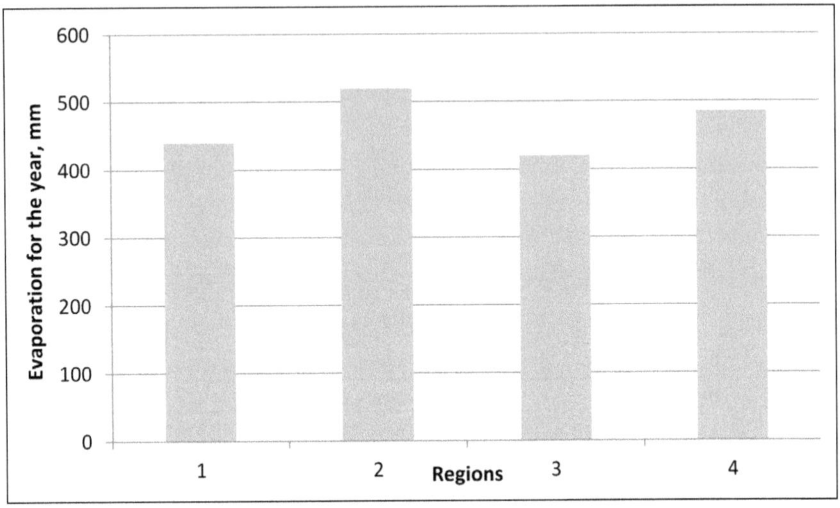

FIGURE 11.6 Evaporation for the year in the regions of Ural (mm). 1—Sverdlovsk; 2—Chelyabinsk; 3—Tyumen; 4—Kurgan.

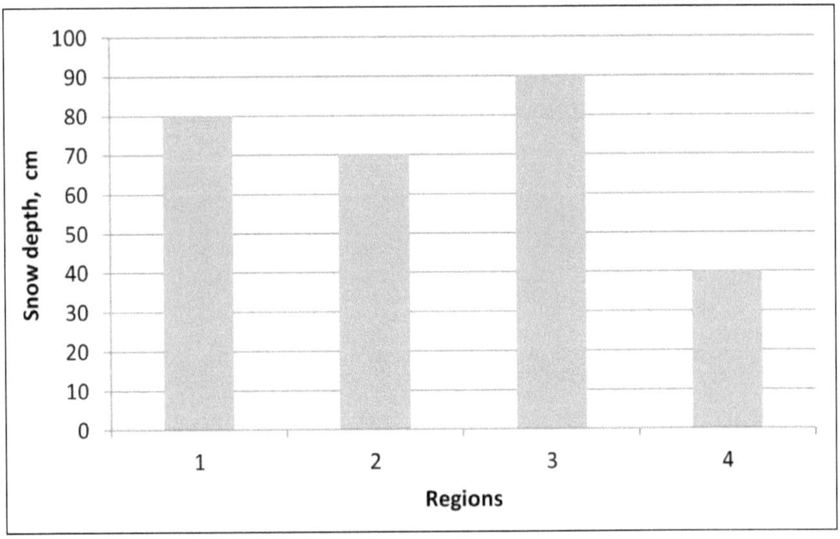

FIGURE 11.7 Snow depth in the regions of Ural (cm). 1—Sverdlovsk; 2—Chelyabinsk; 3—Tyumen; 4—Kurgan.

In the Sverdlovsk region, for the year, from 400 to 700 mm of precipitation falls in the form of rain and snow with evaporation of 375—440 mm. In other words, in some years, evaporation may exceed precipitation, which is fraught with periodic soil and atmospheric droughts, harmful to agriculture. On average, in the soil of this area, the reserve of productive moisture is maintained at the level of 157 mm, that is, higher than in other areas of Trans-Urals. A characteristic feature of the climate of the Trans-Ural region is its continental. Meridional elongated ridge of the Ural mountains is a natural obstacle to Western air flow, deforming the direction of their movement, which affects the climatic zoning of the territory: on the Eastern edge of the Russian plain, bounded from the east by the Ural ridge, the climate is temperate continental, while adjacent to the Ural mountain landscapes of the West Siberian plain, the climate is almost everywhere continental. Thus, the Ural range is a natural boundary between the climatic zones of the European part of Russia and Siberia. With a sufficient amount of temperatures above 10 °C, but a short duration of the frost-free period in the region on summer days, there is often hot weather, up to 30° and above, and strong frosts in winter (Figures 11.1–11.4).

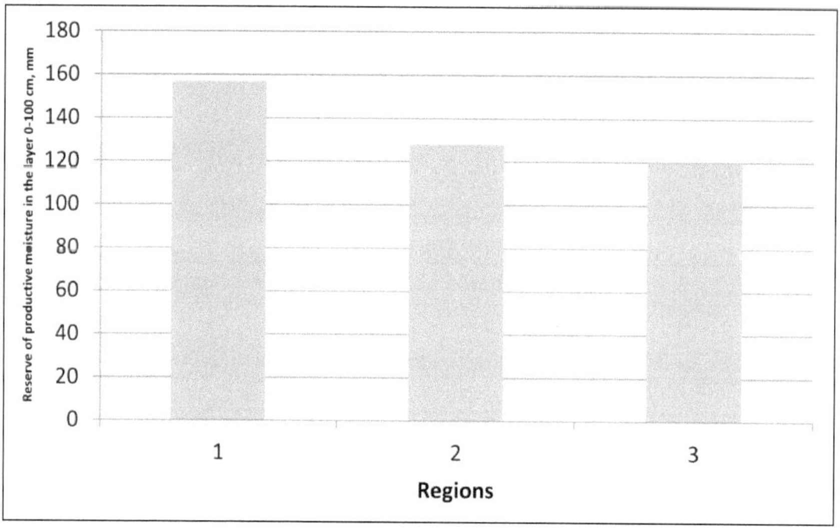

FIGURE 11.8 Reserve of productive moisture in the layer 0–100 cm in spring (mm). 1—Sverdlovsk; 2—Chelyabinsk; 3—Kurgan.

According to the calculations, the hydrothermal index (HI) of weather conditions in these areas, as a rule, is less than one, and only in the Sverdlovsk region, the HI can exceed this value. However, this circumstance does not exclude unfavorable weather conditions of crop cultivation emerging in some years and in the Sverdlovsk region. Weather conditions during 2015—2017 presented in Table 11.4 were taken from the weather station of the Bogdanovichsky District of Kamyshlov City of Sverdlovsk region, where our field experience in fertilizing spring wheat was conducted. Weather conditions during these years of research were different in terms of temperature and moisture content of the soil and fully reflected the features of the Ural region, which had a direct impact on the yield of spring wheat and allowed a comprehensive assessment of the effect of fertilizers used. In 2017, the temperature readings were close to the average annual, and in 2016, they exceeded those in all months of observation. Year 2015 was the most uneven year in terms of heat distribution: in May and June, the recorded air temperature was above normal, followed by a sharp decline without a further increase.

TABLE 11.4 Agrometeorological Conditions During 2015–2017

Amount of Precipitation, mm						
Year	**May**	**June**	**July**	**August**	**September**	**Amount**
2015	82	68	95	101	20	366
2016	10	40	42	36	–	128
2017	32	102	97	75	48	354
Long-term average	41	64	84	59	49	297
Temperature of Air, °C						
Year	**May**	**June**	**July**	**August**	**September**	**Average**
2015	17.4	20.0	16.6	13.7	11.0	15.7
2016	12.6	17.3	19.6	22.1	–	17.9
2017	11.0	16.3	17.6	17.1	11.9	14.8
Long-term average	11.5	16.8	19.2	16.0	10.4	14.8
Hydrothermal Indexes						
Year	**May**	**June**	**July**	**August**	**September**	**Average**
2015	1.6	1.1	1.9	2.5	0.6	1.5
2016	0.3	0.8	0.7	0.5	–	0.6
2017	1.0	2.1	1.8	1.5	1.3	1.5
Long-term average	1.2	1.3	1.5	1.2	1.6	1.4

In 2015 and 2017, the amount of precipitation exceeded the average annual values. Year 2016 is characterized by significantly lower values of precipitation during the growing season. According to the degree of moisture in 2015 and 2017, these years can be characterized as quite humidified. There was a very low value of the HI for all months of 2016, which indicates the acute aridity of this year. It is known that the effectiveness of the use of mineral fertilizers is closely related to the weather conditions of vegetation: in drought conditions, it decreases, and in sufficient moisture conditions, it increases. In years with different precipitation patterns after sowing to caring (May and June), the effect of fertilizers depends on the possibility of their assimilation by plants, which decreases during the spring drought. The smaller the impact of the environment (better conditions), the higher the increase in the yield of spring wheat. In connection with the relatively complex agroecological conditions of agricultural production, the results of studies on improving the technologies of crop cultivation in the region, especially the leading grain crop—spring wheat, including the optimization of mineral nutrition of this crop, are important. This is the subject of the work discussed below.

11.2 METHODS

The field experiment was carried out in 2015–2017 on gray loamy forest soil. The predecessor of spring wheat was black fallow. The soil had, on average, slight reaction $pH_{salt.}$ 5.1 5.3. The number of nutrient mobile forms over the years ranged from 45–50 mg of easily hydrolyzed nitrogen in the first two years of research to 100–115 mg in 2017, of phosphorus 140–200 mg, and of potassium 120–160 mg per kg of soil. The experiment used seeds of spring wheat variety Simbirtsit. The yield of spring wheat in the experiment with increasing doses of each type of fertilizer was studied against the background of two others: nitrogen—on the background of phosphorus and potassium, phosphorus—on the background of nitrogen and potassium, and potassium—on the background of nitrogen and phosphorus. Background variants are control, P60K60, N60K60, and N60P60; the dose of varied element was evenly increased by values from 30 to 120 kg/ha. In the experiment, the yield of grains, the conditions of its formation, as well as the quality indicators of grain products were determined.

11.3 RESULTS

In 2015–2016, variant without fertilizers it is control of Table 11.5 (20.6—2016 and 36.6—2017), the grain yield of spring wheat was 20.6 C/ha, and in 2017 at the site with a higher level of fertility, the yield was 36.6 C/ha (Table 11.5). With an increase in the dose of nitrogen from N30 to N120 against the background of P60K60, the grain yield increased as well as in the drought of 2016 (HI = 0.6) and in the favorable precipitation and temperature year (2015) (HI = 1.5).

TABLE 11.5 Yield of Spring Wheat Depending on Fertilizers

Variant	Yields in Different Years of Research, C/ha				Yield Increase	
	2015	2016	2017	Average for 3 years	C/ha	%
Control	20.6	20.6	36.6	25.9	–	–
Background $P_{60}K_{60}$	26.6	22.6	43.5	30.9	5.0	19.3
$P_{60}K_{60} + N_{30}$	27.8	23.6	42.3	31.2	5.3	20.5
$P_{60}K_{60} + N_{60}$	41.0	23.7	41.7	35.5	9.6	37.1
$P_{60}K_{60} + N_{90}$	42.7	24.5	40.5	35.9	10.0	38.7
$P_{60}K_{60} + N_{120}$	45.2	26.6	38.8	36.9	11.0	42.5
LSD_{05}	1.5	2.3	2.3	–	–	–
Background $N_{60}K_{60}$	29.9	23.9	38.6	30.8	4.9	19.0
$N_{60}K_{60} + P_{30}$	32.9	26.4	39.8	23.0	7.1	27.5
$N_{60}K_{60} + P_{60}$	41.0	23.7	41.7	35.5	9.6	37.1
$N_{60}K_{60} + P_{90}$	42.8	26.1	41.5	36.8	10.9	42.1
$N_{60}K_{60} + P_{120}$	40.5	25.8	40.8	35.7	9.8	37.9
LSD_{05}	1.4	2.1	2.3	–	–	–
Background $N_{60}P_{60}$	34.7	23.4	39.6	32.6	6.7	25.9
$N_{60}P_{60} + K_{30}$	33.6	25.2	38.3	32.4	6.5	25.1
$N_{60}P_{60} + K_{60}$	41.0	23.7	41.7	35.5	9.6	37.1
$N_{60}P_{60} + K_{90}$	37.5	23.1	38.3	33.0	7.1	27.5
$N_{60}P_{60} + K_{120}$	37.1	22.7	41.8	33.9	8.0	30.9
LSD_{05}	1.2	2.9	1.5	–	–	–
Average	35.2	24.2	40.2	33.2	7.3	28.1

Note: for a better perception of the experiment results, the data for variant N60P60K60 are repeated in the table three times, i.e., for each background separately.

The best results are marked in 2015 for variant N120P60K60: It was 70% of the background P60K60. In 2017, the soil of the experimental site was characterized by a higher content of nitrogen and organic matter and the effect of nitrogen fertilizers is expressed implicitly. All variants of experience have provided an increase of yield compared with control. However, with increasing doses of nitrogen, there was a clear trend even of a decrease in the yield from 43.5 C/ha in the variant without nitrogen fertilizers on the background of R60K60 to 38.8 C/ha at the maximum dose of nitrogen. Phosphorus fertilizers in 2017 increased the yield of wheat in doses of more than 30 kg/ha of a.s active substance; d.s.—dry substance (Table 11.7). The impact of potash fertilizers against the background of other fertilizers was not stable due to the high content of this element in the soil. In particular, the maximum doses of K90 and K120 against the background of N60P60 contributed to a decrease in the yield compared to the potassium dose of 60 kg/ha. On average, during the years of research, the highest yield of wheat from increasing doses of potassium of 35.5 C/ha was noted in the N60P60K60 variant. The maximum yields of spring wheat on average for 3 years of 36.8 and 36.9 C/ha were obtained for the variants N60P90K60 and N120P60K60, respectively. The similar values of yield in these variants are explained by the peculiarities of the efficiency of nitrogen and phosphorus fertilizers in 2017 due to the different contents of easily hydrolyzed nitrogen and mobile phosphorus in the soil. In general, within the limits of this experiment, the average yields for the years of research in variants with the use of fertilizers did not go beyond 31–37 C/ha, i.e. the efficiency of mineral fertilizers was characterized by an increase in the yield of 7 C/ha of grain. Fertilizers increased the grain harvest by an average of 28.1% compared to the control. The impact of fertilizers on the yield of spring wheat in the location with high nitrogen content (2017) did not exceed the values in the background. The study of the effect of phosphorus fertilizers on the yield of wheat showed that the highest increase gave a dose of P90 against the background of nitrogen and potassium. Potash fertilizers, applied against the background of nitrogen and phosphorus fertilizers, slightly increased the spring wheat yield. However, their use is necessary, as they accelerate the maturation of crops and increase the resistance of plants to abiotic stress. In general, according to the sum of the indicators, the application of mineral fertilizers in doses of nitrogen, phosphorus, and potassium at 60 kg/ha was the most effective.

It is important to note that mineral fertilizers had a positive impact on the biometric indicators of the emerging crop (Table 11.6). On average, for three years, the mass of 1000 seeds was the maximum for the N60P60K60 variant (44.2 g), and the largest mass of grains was observed in 2015 and 2016 and the lowest in 2017, in which the increased productivity of wheat was obtained. In most variants, there was a negative dependence of the mass of 1000 grains on the grain nature mass. With the largest mass of 1000 grains in the N60P60K60 variant, the smallest grain nature mass was observed in the same variant (765 g/L). Without mineral fertilizers, spring wheat had a nature mass of grains of 732–771 g/L depending on weather conditions: in drought conditions of 2016, it was less, and in temperate years, it was more. In comparison with the control variant, phosphorus fertilizers doses in N60P90K60–N60P120K60 raised the grain nature mass by 17 to 35 g/L and potassium raised it by 12 to 31 g/L. The maximum value of the grain nature mass from the action of nitrogen fertilizers was observed for the variant N90P60K60 in all of the years of studies, where increasing doses of nitrogen did not contribute to the further growth of the nature mass. Although phosphorus fertilizers contributed to the increase in the nature of the grain, to a greater extent than nitrogen and potassium, these fertilizers did not increase the weight of 1000 grains.

The use of mineral fertilizers has a positive impact on the quality of wheat. One of the most important indicators in assessing the quality of grain is its protein content. Mineral fertilizers, especially nitrogen, can significantly affect the change in the content of this indicator. The protein content in the grain of spring wheat without fertilizer (under control) was 10.19–13.18% depending on the external conditions (Table 11.7).

The maximum accumulation of protein (15.75%) was noted in 2016 in the variant with the application of 120 kg/ha nitrogen. This is due to the fact that the period of filling and ripening of grain was accompanied by acute and prolonged drought (for July–August, 1.5–2 times less precipitation compared to the long-term norm). It is established that the increase in the protein content under the influence of adverse factors that inhibit the growth and development of plants is the result of a decrease in the yield due to a decrease in the grain share of carbohydrates, particularly starch. At the same time, the significant costs of nutrients for the development of large vegetative mass and the formation of a high yield with better water availability (2015, 2017) led to a decrease in the protein content in the grain to 12.08%–12.87% at doses of nitrogen N90–N120. The content of gluten, which determines the milling properties of grain, is associated with the level of protein content of grain products. Gluten plays a major role among the factors that determine the technological properties of grain. Gluten, which has high elastic and elastic properties, helps to preserve the consistency of the dough when mixing; this dough is well loosened, keeping its shape. Bread from this dough has a good volumetric yield, shape, and crumb.

TABLE 11.6 Influence of Mineral Fertilizers on the Weight of 1000 Grains and Grain Nature

Variant	Mass of 1000 Grains, g			Grain Nature, g/l		
	2015	2016	2017	2015	2016	2017
Control	42.3	41.0	37.8	742	732	771
Background $P_{60}K_{60}$	44.5	43.5	40.6	746	741	769
$P_{60}K_{60} + N_{30}$	43.0	44.9	40.5	748	746	772
$P_{60}K_{60} + N_{60}$	46.7	45.8	40.1	760	740	765
$P_{60}K_{60} + N_{90}$	47.3	43.9	39.3	761	743	794
$P_{60}K_{60} + N_{120}$	46.8	43.6	39.7	747	735	788
LSD_{05}	0.6	1.3	0.8	13	10	9
Background $N_{60}K_{60}$	43.1	43.6	39.6	747	745	759
$N_{60}K_{60} + P_{30}$	46.5	44.4	39.4	757	744	777
$N_{60}K_{60} + P_{60}$	46.7	45.8	40.1	760	740	765
$N_{60}K_{60} + P_{90}$	46.8	45.8	39.6	776	752	782
$N_{60}K_{60} + P_{120}$	47.3	45.4	39.1	777	748	788
LSD_{05}	0.9	1.6	1.1	11	8	8
Background $N_{60}P_{60}$	43.1	43.1	39.4	756	728	780
$N_{60}P_{60} + K_{30}$	46.4	44.4	39.1	767	740	781
$N_{60}P_{60} + K_{60}$	46.7	45.8	40.1	760	740	765
$N_{60}P_{60} + K_{90}$	46.2	44.6	39.9	771	739	779
$N_{60}P_{60} + K_{120}$		45.3	39.9	773	744	783
LSD_{05}		1.3	0.9	10	14	8

Mineral fertilizers have had a significant impact on the quantity and quality of raw gluten in spring wheat grains. The maximum weights of the wet gluten obtained from the use of mineral fertilizers in doses of N120P60K60, N60P120K60, and N60P60K60 were 27.6%, 27.1%, and 26.7%, respectively. The content of crude gluten in grain reached the maximum values at a dose of mineral fertilizers N120P60K60: 27.5%

crude gluten in conditions of sufficient moisture and 31.1% in the dry period (Table 11.8).

TABLE 11.7 Protein Content in the Grain and Straw of Spring Wheat

Variant	Protein, % d.s. (Grain)			Protein, % d.s. (Straw)		
	2015	2016	2017	2015	2016	2017
Control	10.19	13.18	10.72	2.35	1.94	4.90
Background $P_{60}K_{60}$	10.81	13.60	10.94	2.18	2.39	4.79
$P_{60}K_{60} + N_{30}$	11.24	13.45	11.17	2.97	2.35	4.56
$P_{60}K_{60} + N_{60}$	12.25	13.64	11.86	4.13	2.93	5.13
$P_{60}K_{60} + N_{90}$	11.52	14.57	12.08	3.43	3.14	5.76
$P_{60}K_{60} + N_{120}$	12.87	15.75	11.74	4.25	3.80	5.42
LSD_{05}	0.10	0.89	0.63	0.28	0.26	0.68

TABLE 11.8 Protein Content, and Quantity and Quality of Gluten at Different Doses of Mineral Fertilizers on Average for 2015–2017

Variant	Protein, % d.s.	Gluten, %	Quality of Gluten, units IGD
Control	11.36	24.1	79.3
Background $P_{60}K_{60}$	11.78	24.6	78.0
$P_{60}K_{60} + N_{30}$	11.95	25.4	73.0
$P_{60}K_{60} + N_{60}$	12.58	26.7	69.6
$P_{60}K_{60} + N_{90}$	12.72	26.9	71.5
$P_{60}K_{60} + N_{120}$	13.45	27.6	70.6
Control	11.36	24.1	79.3
Background $N_{60}K_{60}$	12.05	26.8	75.1
$N_{60}K_{60} + P_{30}$	12.23	26.7	70.6
$N_{60}K_{60} + P_{60}$	12.58	26.7	69.6
$N_{60}K_{60} + P_{90}$	12.42	26.5	68.9
$N_{60}K_{60} + P_{120}$	12.76	27.1	69.5
Control	11.36	24.1	79.3
Background $N_{60}P_{60}$	11.83	26.6	75.8
$N_{60}P_{60} + K_{30}$	12.25	25.5	74.0
$N_{60}P_{60} + K_{60}$	12.58	26.7	69.6
$N_{60}P_{60} + K_{90}$	12.41	25.3	67.9
$N_{60}P_{60} + K_{120}$	12.56	24.8	70.8

Note: IGD—The index of gluten deformation.

For a better perception of the results of the experiment, the data for the fertilizer-free variant and the N60P60K60 variant are presented three times in Table 11.8.

According to the content of the mass fraction of crude gluten, wheat grain corresponded to the third class of quality in all years of research. Nitrogen and phosphorus fertilizers contributed to the increase in gluten to a greater extent than potash. High doses of potassium contributed to the reduction of gluten and its quality. Increasing the doses of phosphate and potash fertilizers to 90 kg/ha and above did not have a significant effect on the protein and gluten contents. The reason for the low protein content can be associated with significant precipitation in May and June, which caused intensive nitrogen consumption during the vegetative development of plants and its insufficient supply from the soil during the formation of the crop, as high doses of nitrogen contribute to lodging and delay the maturation of wheat. At the same time, part of the fertilizer is washed out of the root layer. All this can lead to the formation of a crop with a low protein and gluten content. According to the results of three years of research, the removal of nitrogen, phosphorus, and potassium with the yields of the main and byproducts was calculated and the economic balance of these elements was made by taking into account the amount of fertilizers, wheat yield, and their chemical composition, i.e., real experimentally established sources of arrival and consumption (Table 11.9).

The recoupment of fertilizers is most pronounced only in 2015, when the recoupment of N—nitrogen; P—phosphorus; K—potassium (NPK) in relation to the control reached 10—11 kg of grain per kg of fertilizer (Table 11.10).

In arid 2016 and in a safe relation to precipitation and high soil fertility in 2017, the use of mineral fertilizers was low and essentially did not justify, from an economic point of view. The use of mineral fertilizers in 2016 and 2017 was not justified from an economic point of view. In all likelihood, this variant can be recommended for the use of fertilizers in conditions identical to the conditions of our field experience on gray forest soils of Trans-Ural.

The economic efficiency was calculated according to the standard technique, which takes into account the increase in cost of a crop from which the cost of obtaining this gain due to the use of fertilizers is subtracted, and then, we determined the conditionally net income and expected profitability (Table 11.11). The prices of fertilizers and grain products for economic calculations were updated every year due to their instability in the market. The prices for nitrogen fertilizers ranged in years from 11,300 rubles/t in 2015 to 13,000 rubles/t in 2017; for phosphorus, the price was 27,100 rubles/t; and for potassium, the prices ranged from 14,300 rubles/t to 18,500 rubles/t. The price of grains in 2015 amounted to 9500 rubles/t,10,560 rubles/t in 2016, and 8448/t in 2017.

TABLE 11.9 Removal of Nitrogen, Phosphorus, Potassium with the Harvest of Primary and Secondary Products and the Economic Balance of NPK in the Cultivation of Spring Wheat

Variant	Removal of Nitrogen, kg/ha			Nitrogen Balance		
	2015	2016	2017	2015	2016	2017
Control	49.2	57.6	106.6	−49.2	−57.6	−106.6
Background $P_{60}K_{60}$	65.2	64.2	114.3	−65.2	−64.2	−114.3
$P_{60}K_{60} + N_{30}$	73.9	68.8	106.7	−43.9	−38.8	−76.7
$P_{60}K_{60} + N_{60}$	127.7	72.6	112.7	−67.7	−12.6	−52.7
$P_{60}K_{60} + N_{90}$	122.9	86.1	119.5	−32.9	+3.9	−29.5
$P_{60}K_{60} + N_{120}$	151.7	98.9	106.3	−31.7	+21.1	+13.7
Variant	**Removal of Phosphorus, kg/ha**			**Phosphorus Balance**		
	2015	**2016**	**2017**	**2015**	**2016**	**2017**
Control	9.0	6.4	5.4	−9.0	−6.4	−5.4
Background $N_{60}K_{60}$	12.8	7.2	5.4	−12.8	−7.2	−5.4
$N_{60}K_{60} + P_{30}$	16.5	8.3	5.1	+13.5	+21.7	+24.9
$N_{60}K_{60} + P_{60}$	21.1	7.6	5.1	+38.9	+52.4	+54.9
$N_{60}K_{60} + P_{90}$	20.6	8.9	6.0	+69.4	+81.1	+84.0
$N_{60}K_{60} + P_{120}$	20.3	8.6	4.1	+99.7	+111.4	+115.9
Variant	**Removal of Potassium, kg/ha**			**Potassium Balance**		
	2015	**2016**	**2017**	**2015**	**2016**	**2017**
Control	26.2	37.3	33.0	−26.2	−37.3	−33.0
Background $N_{60}P_{60}$	48.2	38.9	38.6	−48.2	−38.9	−38.6
$N_{60}P_{60} + K_{30}$	60.3	54.9	49.5	−30.3	−24.9	−19.5
$N_{60}P_{60} + K_{60}$	68.7	46.7	38.5	−8.7	+13.3	+21.5
$N_{60}P_{60} + K_{90}$	63.7	48.9	50.4	+26.3	+41.1	+39.6
$N_{60}P_{60} + K_{120}$	61.2	54.7	32.7	+58.8	+65.3	+87.3

It is established that on the gray forest soil of the Ural region, you can get the grain yield of spring wheat on an average of 25–26 C/ha with fertilizers, but due to the application of mineral fertilizers—up to

36 C/ha. Spring wheat is most responsive to nitrogen fertilizers applied at doses of up to 60 kg/ha and more on the background of phosphorus and potassium. Phosphorus fertilizers (on the background of nitrogen and potassium) showed maximum efficiency on the yield of spring wheat in doses increasing to P60 and P90. However, already at a dose of P120, there was a decrease in the efficiency of phosphate fertilizers. A similar result can be seen in the effect of potassium fertilizers on the yield of spring wheat. However, the decrease in the effectiveness of potash fertilizers was manifested already at a dose K90. All of these features of the action of mineral fertilizers on the yield of spring wheat are important to consider when developing its fertilizer systems.

TABLE 11.10 Recoupment of Mineral Fertilizers When Applied Under Spring Wheat, kg of grain/kg a.s. of Fertilizer

Variant	Recoupment of NPK in Relation to the Control				Recoupment of NPK in Relation to the Background			
	2015	2016	2017	In General	2015	2016	2017	In General
					Recoupment of N			
Background $P_{60}K_{60}$	5.0	1.7	5.8	4.2	–	–	–	–
$P_{60}K_{60} + N_{30}$	4.8	2.0	3.8	3.5	4.0	3.3	–	2.4
$P_{60}K_{60} + N_{60}$	11.3	1.7	2.8	5.3	24.0	1.8	–	8.6
$P_{60}K_{60} + N_{90}$	10.5	2.4	1.9	4.9	17.9	3.4	–	7.1
$P_{60}K_{60} + N_{120}$	10.2	2.5	0.9	4.5	15.5	3.3	–	6.3
					Recoupment of P_2O_5			
Background $N_{60}K_{60}$	7.8	2.8	1.7	4.1	–	–	–	–
$N_{60}K_{60} + P_{30}$	8.2	3.9	2.1	4.7	10.0	8.3	4.0	7.4
$N_{60}K_{60} + P_{60}$	11.3	1.7	2.8	5.3	18.5	–	5.2	7.9
$N_{60}K_{60} + P_{90}$	10.6	2.6	2.3	5.2	14.3	2.4	3.2	6.6
$N_{60}K_{60} + P_{120}$	8.3	2.2	1.8	4.1	8.8	1.6	1.8	4.1
					Recoupment of K_2O			
Background $N_{60}P_{60}$	11.8	2.3	2.5	5.5	–	–	–	–
$N_{60}P_{60} + K_{30}$	8.7	3.1	1.1	4.3	–	6.0	–	2.0
$N_{60}P_{60} + K_{60}$	11.3	1.7	2.8	5.3	10.5	0.5	3.5	4.8
$N_{60}P_{60} + K_{90}$	8.0	1.2	0.8	3.3	3.1	–	–	1.0
$N_{60}P_{60} + K_{120}$	6.9	0.9	2.2	3.3	2.0	–	1.8	1.3

Note: For a better perception of the results of the experiment, the data for the N60P60K60 variant are presented three times in Table 11.10, that is, for each background separately.

TABLE 11.11 Economic Efficiency of Fertilizer Application on Average Over 2015—2017

Variant	Yield, C/ha	Yield Increase, C/ha	Additional Costs, rubles/ha	Cost of Additional Products, rubles/ha	Shareware Net Income, rubles/ha	Profitability, %	Recoupment of Expenses, rubles/rubles
Control	25.9	–	–	–	–	–	–
Background $P_{60}K_{60}$	30.9	5.0	5612	4547	−1065	−19.5	0.81
$P_{60}K_{60} + N_{30}$	31.2	5.3	6788	14,441	−1847	−27.0	0.73
$P_{60}K_{60} + N_{60}$	35.5	9.6	7963	8987	1024	15	1.15
$P_{60}K_{60} + N_{90}$	36.3	10.4	9139	9892	735	10.9	1.11
$P_{60}K_{60} + N_{120}$	36.9	11.0	10,314	10522	208	5.2	1.05
Background $N_{60}K_{60}$	30.8	4.9	4076	4670	594	22.4	1.20
$N_{60}K_{60} + P_{30}$	33.0	7.1	6019	6838	826	16.9	1.17
$N_{60}K_{60} + P_{60}$	35.5	9.6	7963	8987	1024	15.0	1.15
$N_{60}K_{60} + P_{90}$	36.8	10.9	9907	10,346	439	6.2	1.06
$N_{60}K_{60} + P_{120}$	35.7	9.8	11,869	9315	−2514	−20.1	0.80
Background $N_{60}P_{60}$	32.6	6.7	6238	6295	57	−2.3	1.02
$N_{60}P_{60} + K_{30}$	32.4	6.5	7101	6215	−886	−10.6	0.89
$N_{60}P_{60} + K_{60}$	35.5	9.6	7963	8987	1024	15.0	1.15
$N_{60}P_{60} + K_{90}$	33.0	7.1	8826	6710	−116	−21.5	0.78
$N_{60}P_{60} + K_{120}$	33.9	8.0	9688	7429	−2260	−21.7	0.78

11.4 CONCLUSIONS

All doses of fertilizers studied in the experiment significantly increased the yield of spring wheat relative to control. However, the leading role in increasing the yield of spring wheat in the gray forest soils of the Middle Urals belongs to nitrogen. The maximum yields of the whole experience over the two years of study for variant N120P60K60 amounted to 45.2 (2015) and 26.6 (2016) C/ha. However, on soils with a high level of fertility (nitrogen content), the expected increase in the last year of the experiment laying did not occur; on the contrary, there was a tendency to decrease the yields with increasing doses of nitrogen fertilizers, which is due to the availability of a sufficient supply of nitrogen to plants during the growing season. Normal nitrogen supply also had a significant impact on crop structure (higher straw yield compared to previous years) and plant biometrics (plant height, ear size). Phosphorus fertilizers, on average, gave the maximum yield in the dose of P90 on the background N60K60, and potassium in the dose of K60 on the background N60P60. The use of mineral fertilizers had a positive effect on the quality of wheat grain. Baking properties of grains changed from both the application of mineral fertilizers and weather conditions. Protein accumulation steadily increased with increasing doses of nitrogen fertilizers, but its higher content was observed in the year with insufficient moisture. The use of mineral fertilizers is the most cost-effective in the variant N60P60K60. Potash fertilizers in increasing doses in contrast to nitrogen and phosphorus, to a lesser extent, contributed to the increase in the payment of grain increases, and in less favorable weather conditions in 2016, the amount of grain did not exceed the amount of potassium contributed. In 2017, increasing doses of nitrogen fertilizers has not led to the expected yield increase compared to the background variant P60K60. On average, for three years, the most economically acceptable variant was N60P60K60. In this variant, the obtained relatively high yield of wheat was 35.5 C/ha and the highest net income was 1024 rubles/ha with a profitability of 15% and a payback of 1.15 rubles/rubles. All of the other variants of the field experiment were inferior on the set of economic indicators.

KEYWORDS

- **soil**
- **climate**
- **fertilizers**
- **productivity**
- **quality**
- **income**
- **humus**
- **crop structure**

REFERENCES

Kidin V. V. The system of fertilizer. Moscow: Publishing house of Russian State Agrarian University, Moscow Agricultural Academy, 2012, 534 p. (in Russian).

Koshkin E. N. Physiological resistance of agricultural crops. Moscow: Publishing House «Drofa» ("Bustard in Russian"), 2010, 638 p. (in Russian).

Mineev V. G. Favorites: Collection of scientific articles in 2 parts. Agrochemistry and quality of wheat. Environmental problems and functions of agrochemistry. Moscow: Publishing House of Moscow State University, 2005, 604 p. (in Russian).

Mineev, V. G., Sychev V. G., Gamzikov G. P. et al. Agrochemistry. Mineev V.G. (Ed.). Moscow: Publishing House of Pryanishnikov All-Russian Scientific Research Institute of Agrochemistry, 2017, 854 c (in Russian).

Ministry of Agriculture of Russian Federation: Methodological guidance on the design of fertilizer application in technologies of adaptive landscape agriculture. A. L. Ivanov. L. M. Derzhavin (Eds.). Moscow: Ministry of Agriculture of Russian Federation, 2008, 392 p. (in Russian).

Pryanishnikov D. N. Selected works, Vol. 1. Moscow: «Kolos» ("Ear") Publishing House, 1965, 768 p. (in Russian).

Timiryazev K. A. The life of plants. Moscow: «Novosti» ("News") Publishing House, 2006, 320 p. (in Russian).

INDEX

A

ABE fermentation, 212
Absorption isotherm, 13–14
Adsorbate, 36
Adsorption, 36
Aerodynamic flow. *See* Laminar flow
Agroindustrial wastes
  disposal of waste, 40
  extraction, 53–54
  grape, 46–49
  nopal, 49–53
  nutshell, 40–46
Alkaline cooking, 76
Amylopectin, 208
α-amylose, 208
Anacardic acid, 42
Anarcadium occidentale, 43
Arabic gum, 185
Argentine propolis, 121
Artemia salina, 103
Aspergillus genera, 117–119
Aspergillus versicolor, 118
Atrazine, 114, 121

B

Betaciananias, 167
Betacyanins, 51
Betalains, 51, 162–163
  applications in the food industry, 168–171
  color, 166–167
  microencapsulation of, 171–175
  Myrtillocactus geometrizans, 165
  pharmacological industry, in, 175
  source, 164
  stability of, 171
Beta vulgaris, 164
Betaxanthins, 51, 167
Biotechnological processing, 211
  ABE fermentation, 212
  ethanol, 214
  lactic acid fermentation, 213
  short-chain fatty acid production, 213
  xylitol, 213
Bond length, 24
Botrytis cinerea, 191
Brassica napus, 117
Breakfast cereals, 84
  composition, 85
  extrusion, 85–86
  ingredients and additives, 86
Butylated hydroxyanisole (BHA), 113
Butylated hydroxytoluene (BHT), 113

C

Cashew nut, 43. *See also* Nutshell
Catastrophic nucleation, 18
Cavitation, 44–45
Celiac disease, 62
Cellulose, 208
Cercosporin (CER), 116
Chemical elements, 21
  orbital structures and electronic configurations, 22
  Pauli exclusion principle, 23
  periodic table, 22
Citrinin (CIT), 117–118
Citrus industry, waste of, 156
  biological activity, 167–168
  content of proteins, 159
  graph of, 158
  methods, 156–158
  percentages of dried matter, 158
  values of wet samples, 159
Clostridium acetobutylicum, 212
Clostridium beijerinckii, 212
Colletrotrichum gloeosporioides, 191
Conjugated linolenic acid (CLnA), 136

Control emulsion, 190
Corn, 62–63
  as breakfast cereals, 84–86
  infant food supplements, in, 86–87
  nixtamalization process, 75–76
  nixtamalized products, 75–76
  physicochemical characterization of bakery products based on, 65–66
  as snacks, 77–84
Corn flour (CF), 62
Corn tortillas, 77

D

Darcy's law, 34
Deoxinivalenol, 116
Dietary fiber, 51
Diffusion coefficients, 34
Dimensionless friction factor, 9
Dipolar moment, 27
Dipole–dipole force, 28
DON. *See Deoxinivalenol*
Dosidicus gigas. See Jumbo squid

E

Edible gels, 20
Electrical pulses method, 54
Electroanalytical chemistry
  definition, 111
Electroanalytical methodologies
  electrochemical biosensors, 121
  enzyme-linked immunosorbent assay, 121–122
  linear sweep voltammetry, 122
  mycotoxins, 114–119
  natural antioxidants, 120–121
  sensors, 121
  synthetic antioxidants, 119
Electrochemical biosensors, 112
Electron density, 22
ELISA. *See* Enzyme-linked immunosorbent assay
Emulsion
  analysis of variance, 189
  antioxidant activity, 188
  conjugated dienes in, 188–189
  continuous phase, 15
  destabilization, 16
  determination of stability, 187–188
  dispersal of lipids, 184
  evaluation of solutions, 187
  hydroperoxides, 188–189
  microbial growth, 191
  microbiological assays, 189
  oil-in-water, 15
  opacity and transparency, 187
  stability, 16, 184
  water-in-oil, 15
Encapsulation method, 173–174
Enthalpy, 5–6
Entropy, 4–5
Enzyme-assisted extraction, 53
Enzyme-linked immunosorbent assay, 121–122
Equation of state, 6–7
Equilibrium relative humidity (ERH), 12–13
Escontria chiotilla, 165
Ethanol
  schematic diagram of production, 215, 216
  from stalk juice, 214
  from starch, 214
Expanded snacks, 83–84
Extraction methodologies, 208
  electric pulses, 210
  extrusion, 210–211
  microwave, 209
  ultrasound extraction of bioactive compounds, 209
Extrusion-cooking process, 73–74

F

Flavonoids, 192
Fluorensia cernua, 185
Foams, 16–17
Food and Drug Administration (FDA), 162
Food emulsions, 14
Functional ailments, 156
Functional food, 163
Fusarium Oxysporum, 191
Fusarium proliferatum, 116

G

Gibbs energy, 6
Glicerol, 185
Gluten proteins, 70
Grape, 46–47
  extraction methodologies for bioactive compounds, 47–49
  microwave-assisted extraction, 48–49
  pressurized liquid extraction, 47
  production of, 46
  solid–liquid extraction, 47
  subcritical water extraction, 48

H

Hagen–Poiseuille law, 34
Heisenberg uncertainty principle, 22
Homogeneous volume nucleation, 18
Horseradish peroxidase (HRP) enzyme, 115
Hund rule, 23
Hybrid atomic orbitals, 25
Hydrogen bond, 28
Hydroperoxides, 188–189
Hylocereus undatus, 165. *See* Pitaya
Hypoxanthine, 114

I

Isotopes, 21

J

Jumbo squid, 83

K

Kozeny–Carman equation, 35

L

Lactic acid fermentation, 213
Lactobacillus buchnerii, 213
Laminar flow, 8
  cylinder or tube, in, 8–9
Legumes, 62
Lignocellulose, 208
London dispersion, 28

M

Mammillaria candida Scheidweiler, 165
May pitaya, 165
Microencapsulation, 169
Microwave-assisted extraction, 45–46, 48, 54
Molinate, 114, 121
MON. *See* Moniliformin
Moniliformin, 116
Monoterpene, 114
Mucilage, 51
Mycotoxins, 113
  Alternaria genus, 114–115
  Aspergillus genera, 117–119
  Cercospora genus, 116
  Fusarium fungi, 115–116
  Penicillium genus, 117–119
Myrtillocactus geometrizans, 165

N

Natural antioxidants, 114, 120–121
Natural pigments, 163–166
Newtonian fluids, 8, 10
Nicotinamide adenine dinucleotide phosphate (NADP), 138
Nilson's equation, 167
Nixtamal, 77
Nixtamalized corn products, 75–76
  chips, 79–80
  expanded snacks, 83–84
  mixtures of nixtamalized cornflour and legumes, snacks from, 81–83
  nixtamalization process, 76–77
  snacks, 77–79
Noble gases, 23
Nopal, bioactive compounds from, 49–50
  cladodes, 50–51
  extraction methods, 53–54
  flowers, 51–52
  fruit peel, 52
  fruit (tuna), 51
  residues, 52
  thorns, 52
Nucleation, 17
  crystallization of colloids, 17–18
  food colloids, in, 18–19
  types of, 18

Nutraceuticals, 142–144, 163
Nutshell, 41
carbon source, 41
components, 42
disposal, 40
extraction methodologies for bioactive compounds, 43–46
phenolic compounds, extraction of, 42

O

Ochratoxin A (OTA), 117
Octet rule, 23
Oil-in-water (O/W) emulsion, 185
Oncorhynchus mykiss, 146
Opuntia engelmanni, 52
Opuntia ficus-indica, 175
Opuntia lasiacantha, 169
Orbitals, 22
Organic compounds, 24
attraction forces, kinds of, 28
covalent bond, 27
hybrid atomic orbitals, 25
hybridized orbitals, 26
molecular bonds in, 25
valence shell electron pair repulsion theory, 24

P

Pasta
as artificial food, 71
physicochemical properties, 66–69
Patulin (PAT), 118
Pauling electronegativity, 27
Pectin, 51
Penicillium, 117–119
Phenolic antioxidants, 113
Physical chemistry
application of, 2–3
definition of, 2
Physicochemical properties of
bakery products, 63–64
based on corn and legumes, 64–66
mixtures of corn and legumes, pasta made from, 69–75
pasta, 66–69
rambutan peel, 98–99
rambutan seed, 99–100
rambutan seed fat, 100–102
rambutan seed starch, 103
Pitaya, 165
Poisson's ratio, 21
Polar covalent bond, 27
Polyphenolic biomolecules, 206
Pomegranate, 130
applications, 144–146
aril, 132
chemical composition, 131
constituents, 131–133
peel, 132–133
producers of, 130
scientific perspectives, 147
seed oil, 133–136
seeds, 133
Pomegranate seed oil, 133–136
applications, 144–146
effect of extraction method, 139–140
extraction methods, 138–142
fatty acid composition of, 135
scientific perspectives, 147
Potential pulse voltammetric techniques, 111–112
Power law model, 32
Prickly pear *(O. ficus-indica),* 49–50
Propolis, 114
Propyl gallate (PG), 113
Punica granatum. See Pomegranate
Punica granatum (PgFac), 137
Punicic acid (PuA)
chemical structure, 136
linoleic acid and, 137
metabolism, 136–138
natural biosynthesis, 136–138

Q

Quantum chemistry
atomic structure, principles of, 21–23
chemical bonds, 23–24
molecular orbitals, 24–25
orbital structures and electronic configurations, 22
polarity and molecular interactions, 25–28

R

Rambutan, physicochemical properties, 96–97
  peel, 98–99
  pulp, 97–98
  seed, 99–100
  seed fat, 100–101
  seed starch, 103–105
  volatile odor compounds in fruit, 99
Rambutan seed fat
  fatty acids in, 101
  physicochemical properties of, 100–102
  pretreatments, 102–103
  triacylglycerols present in, 101
Rambutan seed starch
  compared to corn starch, 104
  physicochemical properties, 103
  pretreatments to improve, 104–105
  with and without alkaline treatment, 105
Reaction order, 32
Ready-to-eat (RTE) products, 78
Resveratrol, 207
Reynolds (Re) number, 8
Rheology
  definition, 7
  solids, 21
Rhizopus stolonifer, 191
RTE cereals, 84

S

Sea of electrons, 23
Semolina pasta, 70
Soft solids
  definitiion, 19
  edible gels, 20
  gel, 19–20
Sorghum, 200–202
  bioactive compound extraction methods, 208–211
  bioprocessing, 211–214
  dietary fiber in, 203
  flavonoids, 205
  lignocellulosic biomass, 208
  mineral content, 203–204
  nutritive value, 202–203
  phenolic acids, 205
  polyphenolic biomolecules, 206
  stilbenes, 207
  useful components of, 204–205
Sorghum bicolor. See Sorghum
Sorption isotherm, 13–14
Soxhlet extraction, 44
Spring wheat in Ural region, 224. *See also* Ural region, climatic conditions of
  duration of frost-free period, 225
  economic efficiency, 237–238, 240
  humus contents in soil, 227
  influence of mineral fertilizers, 235
  maximum temperatures, 224–226
  protein content in, 236
  recoupment of mineral fertilizers, 239
  removal of, 238
  yield of, 232–235
Starch, 208
Stenocereus griseus, 165
Sterigmatocystin (STEH), 118–119
Stoichiometric coefficient, 28
Stoichiometry theory, basics of, 30
Stripping voltammetry, 112
Subcritical water extraction, 48
Supercritical fluid extraction, 54
Surface of interfacial free energy, 35
Surface tensions, 35–36
  solids and liquids, between, 36
  surface of interfacial free energy, 36
Surfactants, 36
Synthetic antioxidants, 113–114, 119
Synthetic dyes, 164

T

Tannins, 192
Tarbush, 185
  antioxidant activity, 193
  biological activity of, 185
  effect of the extract, on Candelilla Wax-Based Emulsion, 190, 192
  evaluation of emulsion solutions, 187
  preparation of emulsion, 186–187
  vegetal material, 186
Tert-butyl hydroxyquinone (TBHQ), 113–114
Thermodynamics
  definition, 3–4

principles of, 3
thermal property, 4
Tocopherols, 114
Transport phenomena, 7, 34
Trichosantheskirilowii (TkFac), 137
Turbulent flow, 8, 9
cylinder or tube, in, 9–10

U

Ultramicroelectrodes (UMEs), 112
Ultrasound-assisted extraction (UAE), 44–45, 54
Ural region, climatic conditions of, 224
accumulation of protein, 234
acid soils in, 226
agrometeorological conditions during 2015–2017, 230
annual rainfall in, 227
area of arable land, 224
contents of mobile potassium and phosphorus, 227
duration of frost-free period, 225
evaporation, 228
field experiment, 231
humus, 227
influence of mineral fertilizers, 235
maximum temperatures, 224
reserve of productive moisture in, 229
snow depth, 228
temperature of the coldest month, 226
temperature of the warmest month, 225
yield of spring wheat, 232

V

Valence shell electron pair repulsion theory, 24
van der Waals forces, 27, 29
Viscosity
definition, 10
function, 11
Vitis vinífera. See Grape
Volatile compounds, 23
Vulgaxanthin, 166

W

Water activity, 12–13
Water content
in vegetables and fruits, 11–12
Wheat, 62
Wheat flour (WF), 62

X

Xanthine, 114
Xylitol, 213

Y

Yersinia ruckeri, 146

Z

ZEA. *See Zearalenone*
Zea mays. *See* Corn
Zearalenone, 115